Application of Expert Systems - Theoretical and Practical Aspects

Edited by Ivan Nunes da Silva and Rogério Andrade Flauzino

Published in London, United Kingdom

IntechOpen

Supporting open minds since 2005

Application of Expert Systems - Theoretical and Practical Aspects
http://dx.doi.org/10.5772/intechopen.85202
Edited by Ivan Nunes da Silva and Rogério Andrade Flauzino

Contributors
Andressa Nicolau, Marcelo Carvalho Dos Santos, Victor Henrique Cabral Pinheiro, Filipe Santana Moreira Do Desterro, Roberto Schirru, Mario Campos, Alexandre Wanderley, Ivan Nunes da Silva, Danilo Spatti, Luisa Liboni, Marcel Araujo, Renato Bossolan, Bruno Vitti, João Paulo Lemos Escola, Rodrigo Capobianco Guido, Alexandre Moraes Cardoso, Douglas Maccagnan, João Marcelo Ribeiro, José Ricardo Ferreira Cardoso, Juan Ignacio Guerrero Alonso, Antonio Martín, Enrique Personal, Antonio Parejo, Carlos Leon, Brahim Lejdel, Gonzalo Astray, Manuel Alonso Ferrer, Juan C. Mejuto, Rogério Andrade Flauzino, Bruno Augusto Trevisam

First published in London, United Kingdom, 2020 by IntechOpen
IntechOpen is the global imprint of INTECHOPEN LIMITED, registered in England and Wales, registration number: 11086078, 7th floor, 10 Lower Thames Street, London, EC3R 6AF, United Kingdom
Printed in Croatia

British Library Cataloguing-in-Publication Data
A catalogue record for this book is available from the British Library

Additional hard and PDF copies can be obtained from orders@intechopen.com

Application of Expert Systems - Theoretical and Practical Aspects
Edited by Ivan Nunes da Silva and Rogério Andrade Flauzino
p. cm.
Print ISBN 978-1-83881-006-1
Online ISBN 978-1-83881-007-8
eBook (PDF) ISBN 978-1-83881-008-5

Meet the editors

Ivan Nunes da Silva was born in São José do Rio Preto, Brazil, in 1967. He graduated in computer science and electrical engineering from the Federal University of Uberlândia, Brazil, in 1991 and 1992, respectively. He received both an MSc and PhD in Electrical Engineering from University of Campinas (UNICAMP), Brazil, in 1995 and 1997, respectively. Currently, he is Full Professor at the University of São Paulo (USP). His research interests are within the fields of power system and computational intelligence. He is also associate editor of the *International Journal on Power System Optimization* and editor-in-chief of the *Journal of Control, Automation and Electrical Systems.* He has published more than 400 papers in congress proceedings and international journals and authored several book chapters.

Rogerio Andrade Flauzino was born in Franca, Brazil, in 1978. He received an undergraduate degree and an MSc in Electrical Engineering from São Paulo State University (UNESP), Brazil, in 2001 and 2004, respectively. He received a PhD in Electrical Engineering from the University of São Paulo (USP), Brazil, in 2007. Currently, he is Associate Professor at the University of São Paulo. His research interests include artificial neural networks, computational intelligence, fuzzy inference systems, and power systems.

Contents

Preface XI

Section 1
Application in Support Decision Systems 1

Chapter 1 3
A Case Study of Wavelets and SVM Application in Coffee Agriculture: Detecting Cicadas Based on Their Acoustic and Image Patterns
by João Paulo Lemos Escola, Rodrigo Capobianco Guido, Alexandre Moraes Cardoso, Douglas Henrique Bottura Maccagnan, João Marcelo Ribeiro and José Ricardo Ferreira Cardoso

Chapter 2 15
Artificial Intelligence Models to Predict the Influence of Linear and Cyclic Polyethers on the Electric Percolation of Microemulsions
by Manuel Alonso-Ferrer, Gonzalo Astray Dopazo and Juan Carlos Mejuto

Chapter 3 27
Increasing the Efficiency of Rule-Based Expert Systems Applied on Heterogeneous Data Sources
by Juan Ignacio Guerrero Alonso, Enrique Personal, Antonio Parejo, S. García, Antonio Martín and Carlos León

Chapter 4 37
A Conceptual Framework for Modeling Smart Parking
by Brahim Lejdel

Chapter 5 53
Alert Diagnostic System: SDA
by Andressa dos Santos Nicolau, Marcelo Carvalho dos Santos, Victor Henrique Cabral Pinheiro, Filipe Santana Moreira do Desterro, Roberto Schirru, Mario Cesar Mello Massa de Campos and Alexandre Wanderley

Section 2
Application in Electrical Power Systems 67

Chapter 6 69
Intelligent System for the Estimation of Gases Dissolved in Insulating Mineral Oil from Physicochemical Tests
by Ivan Nunes da Silva, Rogério Andrade Flauzino, Danilo Hernane Spatti, Renato Pagotto Bossolan and Bruno Augusto Trevisam

Chapter 7 85
Efficient Asset Management Practices for Power Systems Using Expert Systems
by Danilo Spatti, Luisa H.B. Liboni, Marcel Araújo, Renato Bossolan and Bruno Vitti

Chapter 8 101
Computational Intelligence to Estimate Fault Rates in Power Transformers
by Danilo Spatti, Luisa H.B. Liboni, Marcel Araújo, Renato Bossolan and Bruno Vitti

Preface

What are expert systems? What are their purposes? What are the impacts resulting from their implementations?

This book aims to answer these questions and more. Written by experts in the field, chapters cover various concepts related to expert systems, including computational intelligence, signal processing, real time systems, systems optimization, electric power systems, support vector machines, fault diagnosis, asset management, and smart cities. Chapters systematically discuss the thematic concepts of expert systems and present a broad range of technical and theoretical information. As such, in addition to being of interest to a professional audience, the book is useful as a text for undergraduate and graduate courses. The prerequisites for understanding this book's content are basic, requiring only elementary knowledge of computation.

The first part of this book (Chaps. 1 – 5) examines the various uses and applications of expert systems. These include support decision systems, alert diagnosis systems, rule-based expert systems applied on heterogeneous data sources, detection of insects in coffee agriculture, influence of linear/cyclic polyethylenes on the electric percolation of microemulsions, and proposition of multi-agent for modeling smart parking.

The second part of this book (Chaps. 5 – 8) presents solutions that comprise technologies of expert systems in electric power systems. It describes topics related to efficient asset management practices for power systems using expert procedures, intelligent systems for estimation of gases dissolved in insulating mineral oil from physicochemical tests, and systems based on computational intelligence to estimate failure rates in power transformers.

Ivan Nunes da Silva and Rogério Andrade Flauzino
University of São Paulo,
Brazil

Section 1

Application in Support Decision Systems

Chapter 1

A Case Study of Wavelets and SVM Application in Coffee Agriculture: Detecting Cicadas Based on Their Acoustic and Image Patterns

João Paulo Lemos Escola, Rodrigo Capobianco Guido, Alexandre Moraes Cardoso, Douglas Henrique Bottura Maccagnan, João Marcelo Ribeiro and José Ricardo Ferreira Cardoso

Abstract

One of the main problems in agriculture is crop pest management, which causes financial damage to farmers. This management is traditionally performed with pesticides; however, with a large area of application, it would be more economically viable and more environmentally recommended to know precisely the regions where there is concrete infestation. In coffee farms, cicada makes a distinctive sound when it hatches after years of underground nymph-shaped living. One possibility of contributing to its management would be the development of a device capable of capturing the sound of the adult cicada in order to detect its presence and to quantify crop insects. This device would be spread across the coffee plots to capture sounds within the widest possible area coverage. With monitoring and quantification data, the manager would have more input for decision-making and could adopt the most appropriate management technique based on concrete information on population density separated by crop region. Thus, this chapter presents an algorithm based on wavelets and support vector machines (SVMs), to detect acoustic patterns in plantations, advising on the presence of cicadas.

Keywords: acoustic patterns, support vector machines, wavelets, digital signal processing, cicada

1. Introduction

Since human population has been intensively increasing, the need for food and other products from agricultural fields also grows. To guarantee the production, monoculture is carried out at extensive areas, which intensifies the appearance of pests. Generally, pest control in agriculture is performed with the use of chemical pesticides, often applied even in areas without pest incidence, which raises the cost of production and may cause environmental impacts that affect human health.

Figure 1.
Quesada gigas. *On the left, male emitting acoustic signals. On the right, lateral view of male resting.*

The development of specific hardware and software for pest detection in agriculture can provide support for distinct production forms which reduce negative impacts. In this sense, the capture, recording, and analysis of acoustic signals emitted by insects can be an alternative to optimize the production of certain crops [1].

The cicada (Hemiptera: Cicadidae) is a good example of an insect capable of emitting acoustic signals. In Brazil, the coffee plantation can be attacked by several arthropods and among them, *Quesada gigas* is considered a key pest in the entire state of Minas Gerais and in the northeastern region of the state of São Paulo [2]. Considering that there have been reports on the occurrence of cicadas in coffee plantations from the period between 1900 and 1904, it is important to say that this fact has interfered in the way of conducting the crop: it has practically forced coffee growers to adopt practices for productive system improvement, such as larger spaces between plants, allowing for the mechanization of crop aiming at the use of pesticides to manage this and other pests and diseases that affect productivity.

Cicadas attack crops in the search for sap, their main food. The impacts on the plants occur in the nymphal phase of the cicada when it sucks sap from the host plant root [3]. The *Q. gigas* species, that are the biggest in size the country, can reach 70 mm in length, including wings, and 20 mm in width in the case of males. The females can reach 69 mm in total length and 16.5 mm in width, as shown in **Figure 1**. The size of the insect is probably the reason why it is associated with the impact caused in coffee plantation.

Males usually sound from October to December. In the 1970s, since there was no efficient method for cicadas controlling, many coffee growers had no choice than to eradicate infested crops. Many of them even abandoned their cultivation. However, lately the recommended control has been made through systemic chemical defenses and more recently by the use of a sound trap that attracted *Q. gigas* to a closed spraying system [4].

There are few technological devices used in coffee plantations to monitor and control the cicadas, and maybe, because of this, the most used device is the chemical one. Considering that the methods currently used for mapping and monitoring the populations of cicadas consist basically of the nonautomated counting of the individuals through direct observation, this paper aims to present some applications using digital signal processing and support vector machines (SVMs) as techniques for detecting and monitoring cicadas in crops and forests, reducing time and control costs.

2. Initial considerations

Machine learning is a branch of artificial intelligence that seeks to develop algorithms capable of learning certain behaviors or patterns through examples, being able to generalize from training. SVMs, for instance, are supervised machine learning methods with superior results compared to other pattern classification procedures, considering binary problems [5].

Thus, one possibility for improving agricultural procedures would be the development of a device capable of capturing the sound of the adult cicada in order to detect its presence, thus monitoring and quantifying crop insects. This device would be spread across the coffee fields to capture sounds within the widest possible area coverage. That allows for the manager to have more input for decision-making, adopting the most appropriate management technique based on concrete information on population density separated by crop region.

Cicadas emit particular frequency components which characterize certain patterns. To detect them, both Fourier and discrete wavelet (DWT) transforms, which convert a time-domain signal to the frequency domain, can be used. Nevertheless, in the latter case, it is still possible to obtain the time support of frequencies [6]. Then, an SVM can be used to refine the results, pointing out the existence of the important patterns from the wavelet-transformed signals. This is just how the proposed approach was implemented. Related works, such as [1, 7–15], perform automated data collection for monitoring and corroborate the present work.

3. Application one (AP1): cicada density estimation by audio processing

This solution consists of a system that, from an input *wav* audio file, discriminates between three possibilities, noise, low density, and high density, assisting in monitoring the cicada infestation in the coffee crop. Based on features from the human auditory system, which easily differentiates between these acoustic patterns, the system proves to be efficient. Working similarly to the cochlea in the human ear [16] and based on the DWT packet, an efficient time-frequency mapping [6] is provided.

Based on our assumptions and intending to assess it, we carried out the following preprocessing procedure to convert each acoustic input signal of variable length into a 25-sample long feature vector:

- AP1 PRE-PROCESSING PHASE:
  - BEGINNING.
  - STEP 1: the raw data from the input signal i, recorded as a *wav* file sampled at 44,100 samples/s, 16-bit [16], is extracted and stored as the vector $s_i[\cdot]$, for $(0 \leqslant i \leqslant X-1)$. Each original *wav* file lasts about 3 s, i.e., $44100 \cdot 3 = 132300$ samples, with variations among them. Since a wavelet-based transformation is used in the next step, we cut the the vectors $s_i[\cdot]$ taking advantage of their central part in such a way that their length became 262,144 samples, which is a power of two;
  - STEP 2: $s_i[\cdot]$ of size 262,144 is converted into its corresponding feature vector, i.e., $f_i[\cdot]$ of size $X = 25$, where $(0 \leqslant i \leqslant X-1)$. Particularly, $f_i[j]$, i.e., the jth component of $f_i[\cdot]$ where $(0 \leqslant j \leqslant 24)$, corresponds to the normalized energy of the jth Bark scale band of the wavelet-packet transformed input signal $s_i[\cdot]$ at the maximum decomposition level, as in Eq. 1, considering the natural frequency ordering [6], according to **Table 1**. Comments on the wavelet family used are presented ahead.

- END.

$$\frac{log\,(input\_size)}{log\,(2)} = max\_level \tag{1}$$

We apply here the technique described in [17], which is based on paraconsistent logic, to analyze the behavior and suitability from the obtained data, i.e., the feature

Bark[a]	Band range (Hz Hz)	Initial WPT sample	Final WPT sample	Energy range[b]
0	0–100	0	$\lfloor \frac{100}{0.0841} \rfloor = 1189$	0–99.9949
1	100–200	1190	$\lfloor \frac{200}{0.0841} \rfloor = 2378$	100.0790–199.9898
2	200–300	2379	$\lfloor \frac{300}{0.0841} \rfloor = 3567$	200.0739–299.9847
3	300–400	3568	$\lfloor \frac{400}{0.0841} \rfloor = 4756$	300.0688–399.9796
4	400–510	4757	$\lfloor \frac{510}{0.0841} \rfloor = 6064$	400.0637–509.9824
5	510–630	6065	$\lfloor \frac{630}{0.0841} \rfloor = 7491$	510.0605–629.9931
6	630–770	7492	$\lfloor \frac{770}{0.0841} \rfloor = 9155$	630.0772–769.9355
7	770–920	9156	$\lfloor \frac{920}{0.0841} \rfloor = 10939$	770.0196–919.9699
8	920–1080	10,940	$\lfloor \frac{1080}{0.0841} \rfloor = 12841$	920.054–1079.9281
9	1080–1270	12,842	$\lfloor \frac{1270}{0.0841} \rfloor = 15101$	1080.0122–1269.9941
10	1270–1480	15,102	$\lfloor \frac{1480}{0.0841} \rfloor = 17598$	1270.0782–1479.9918
11	1480–1720	17,599	$\lfloor \frac{1720}{0.0841} \rfloor = 20451$	1480.0759–1719.9291
12	1720–2000	20,452	$\lfloor \frac{2000}{0.0841} \rfloor = 23781$	1720.0132–1999.9821
13	2000–2320	23,782	$\lfloor \frac{2320}{0.0841} \rfloor = 27586$	2000.0662–2319.9826
14	2320–2700	27,587	$\lfloor \frac{2700}{0.0841} \rfloor = 32104$	2320.0667–2699.9464
15	2700–3150	32,105	$\lfloor \frac{3150}{0.0841} \rfloor = 37455$	2700.0305–3149.9655
16	3150–3700	37,456	$\lfloor \frac{3700}{0.0841} \rfloor = 43995$	3150.0496–3699.9795
17	3700–4400	43,996	$\lfloor \frac{4400}{0.0841} \rfloor = 52318$	3700.0636–4399.9438
18	4400–5300	52,319	$\lfloor \frac{5300}{0.0841} \rfloor = 63020$	4400.0279–5299.9820
19	5300–6400	63,021	$\lfloor \frac{6400}{0.0841} \rfloor = 76099$	5300.0661–6399.9259
20	6400–7700	76,100	$\lfloor \frac{7700}{0.0841} \rfloor = 91557$	6400.0100–7699.9437
21	7700–9500	91,558	$\lfloor \frac{9500}{0.0841} \rfloor = 112960$	7700.0278–9499.9360
22	9500–12,000	112,961	$\lfloor \frac{12000}{0.0841} \rfloor = 142687$	9500.0201–11999.9767
23	12,000–15,500	142,688	$\lfloor \frac{15500}{0.0841} \rfloor = 184304$	12000.0608–15499.9664
24	15,500–22,050	184,305	$\lfloor \frac{22050}{0.0841} \rfloor = 262143$	15500.0505–22050.0000

[a] *Bark band.*

[b] *Range in f_i[Bark band] Hz.*

At that level, which is the maximum, and considering the original signal sampling rate of 44,100 samples/s, each sample of the transformed signal has a resolution of $\frac{44,100}{2^{18+1}} = 0.0841$ Hz. *The energy of each one of the 25 sets is calculated separately and, then, normalized based on its division by the total energy. Thus,* $f_i[j]$ *is the jth normalized energy,* $(0 \leqslant j \leqslant 24)$*, for a certain input signal* $s_i[\cdot]$*. The symbol* $\lfloor \cdot \rfloor$ *in the fourth column of the table represents a rounding floor operation. It is required because a sample is obviously always an integer number. Due to that rounding, the frequency range we obtained from the WPT tree is only an approximation to the Bark scale; however, it does not make any difference in practice.*

Table 1.
The 25 sets of samples, from the 18-th level WPT of size 262,144, is used to mimic the bark scale.

vector. Thus, the data vectors were represented in the paraconsistent plane as point P, according to **Figure 2**. Ideally, the closer the P is from corner $(G_1, G_2) = (0, 1)$, the better our feature vectors separate between the classes, disregarding any specific classifier. The next step was to choose the best wavelet family, that is, the family that puts P closer to that corner.

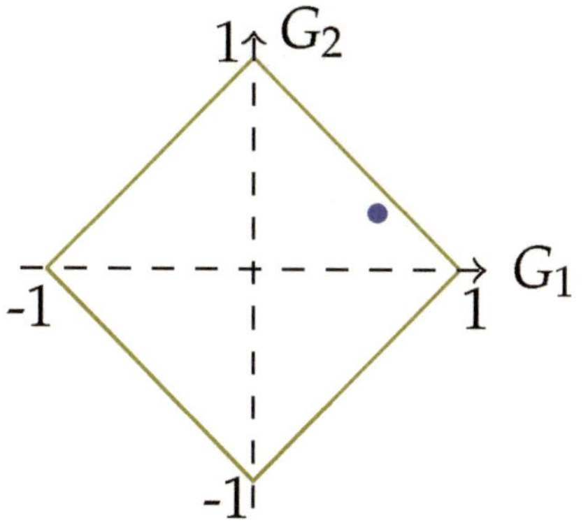

Figure 2.
The paraconsistent plane where the axes G_1 and G_2 represent the degrees of certainty and contradiction, respectively. $P = (G_1, G_2) = (\alpha - \beta, \alpha + \beta - 1)$, drawn in blue just to exemplify, is an important element for our analysis: The closer it is to the corner (1,0), the weaker the classifier associated with the features vector can be. The values of α and β are derived from intra-class and inter-class analyses, respectively, as detailed in [17].

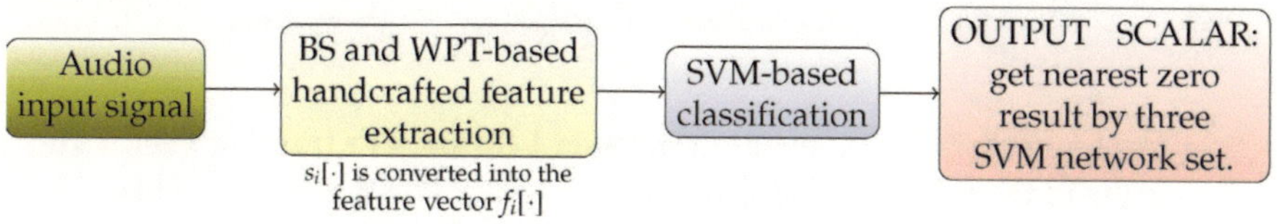

Figure 3.
The experimental setup for the proposed application one.

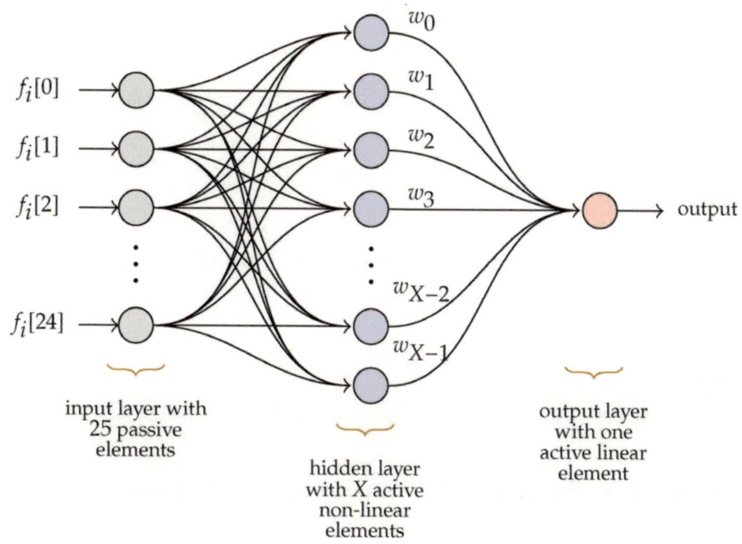

Figure 4.
The SVM structure used in application one approach. The weights determined during the supervised part of the training are $\{w_0, w_1, ..., w_{X-1}\}$. The output element linearly combines the outputs of the hidden layer with the weights.

Proceeding, we estimate that an SVM could be a proper classifier to interpret the feature vectors we selected during the preprocessing step, because of its excellence in terms of binary classifications [5]. Accordingly, **Figure 3** shows the complete setup for the proposed approach. It is divided into two phases, i.e., training and testing, with four steps each. As discussed ahead, a total of X and Y vectors were isolated to carry out each phase, respectively, where $X + Y$ corresponds to the number of acoustic files in the database.

The SVM has been implemented, as described in [5], in such a way that it receives the input vectors defined in the preprocessing step. In **Figure 3**, we illustrate the proposed setup, which is divided into two phases, training and testing, with four steps each. As we will discuss later, a total of X and Y vectors have been isolated to perform each phase, respectively, where $X + Y$ corresponds to the number of acoustic files in the database.

The detailed procedures are as follows:

- AP1 TRAINING PHASE:
  - BEGINNING
  - All the X vectors $f_i[\cdot]$ were used to train an SVM with 25 input passive elements, X hidden active non-linear elements and one output active linear element, as in **Figure 4**. X elements were used in the hidden layer to allow for a simple and effective training scheme, as explained in [5]. The kth element in the hidden layer uses a function of the form

$$e^{-\text{Euclidian\_Distance}\left(k-\text{th\_training\_vector},\text{input\_vector\_under\_analysis}\right)},$$

  - implying that the kth element outputs 1 for the kth training vector and a value in the range (0–1) for the others, where $(0 \leqslant k \leqslant X-1)$. This corresponds to a non-supervised procedure. There is no weight between the input and the hidden layers, however, there are X between the hidden and the output layer. To find them, a linear system of X equations in X unknowns is established and solved, implying in a supervised task. In that system, the closest resultant value from the SVM set corresponds the answer.
  - END.

Once the training procedures are over, the system is ready for testing, as follows.

- AP1 TESTING PHASE:
  - BEGINNING.
  - Each one of the Y testing vectors of size 25 are passed through the trained SVM and the corresponding output is verified: SVM with result closest to zero will be elected;
  - END.

4. Application two (AP2): cicada density estimation by image processing

Visually, cicadas are quite noticeable in the farming environment, so a management hypothesis would be the inclusion of a camera to permit visual detection of pests, adjusting the data capture interval for sending to a web server.

The tests were performed by using images captured in the coffee crop and divided into three classes that represent the pest insidious density: high, medium, and none. The submitted images have been converted to grayscale and have a size of 320 × 240 pixels, as in **Figure 5**.

Similar to AP1, an SVM is used to classify the preprocessed input energies; however, in this case we use the N normalized energies of each wavelet sub-band, instead of 25 Bark bands, according to the decomposition level defined in each test instance. In **Figure 6**, we present a flowchart that illustrates the process of capturing and processing the images.

In **Figure 7**, we have the SVM illustration with $n = 4^{level}$ entries, corresponding to the number of energies of the selected level.

Figure 5.
Examples of images used: high, low, and zero density, respectively.

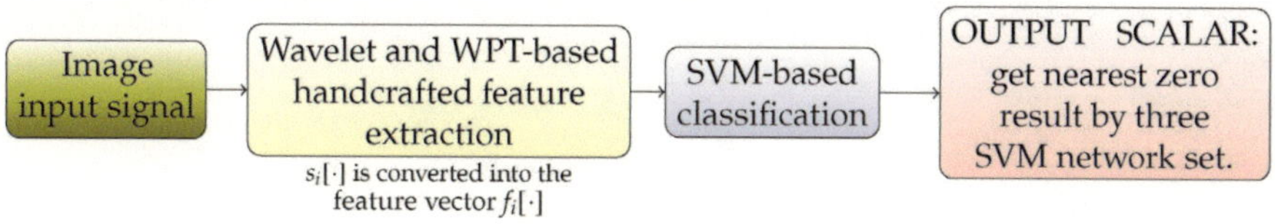

Figure 6.
The experimental setup for the proposed application two.

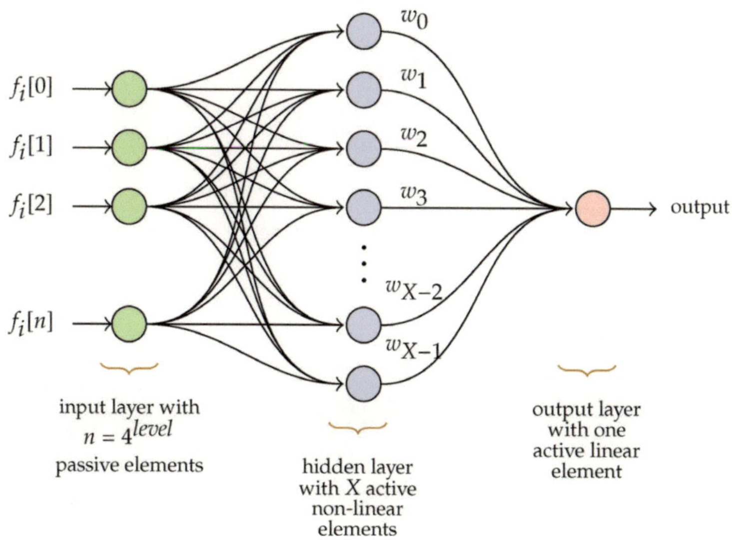

Figure 7.
The SVM structure used in application two. Similar to AP1, the weights determined during the supervised part of the training are $\{w_0, w_1, \dots, w_{X-1}\}$*. The output element linearly combines the outputs of the hidden layer with the weights.*

5. Tests and results

The implementation of the algorithms proposed here were performed using Java programming language.

Wavelet	Train files/class	Test files	Percentage
DAUB6	2	99	94.95/100
DAUB8	3	96	96.88/100
DAUB4	4	93	96.77/100
DAUB4	5	90	96.67/100

Table 2.
Test results from AP1.

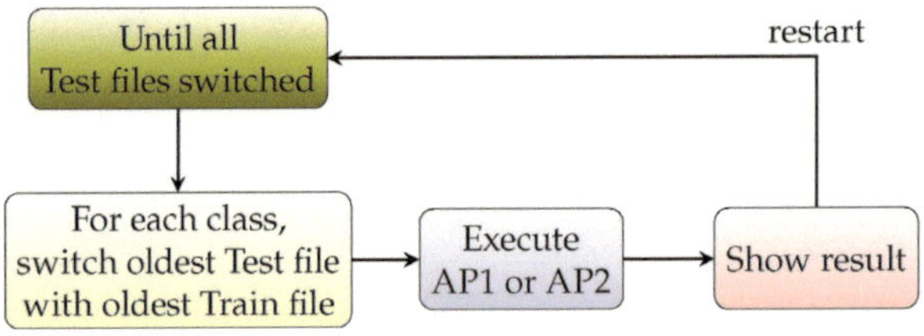

Figure 8.
Cross-validation algorithm.

Wavelet	Level	Train files/class	Test files	Percentage
COIF6	1	2	99	68.69/100
COIF6	1	3	96	81.25/100
DAUB68	1	4	93	81.72/100
DAUB32	1	5	90	83.33/100
DAUB12	2	2	99	86.87/100
DAUB52	2	3	96	87.50/100
DAUB16	2	4	93	87.10/100
DAUB38	2	5	90	87.78/100
DAUB18	3	2	99	86.87/100
DAUB68	3	3	96	89.58/100
DAUB24	3	4	93	90.32/100
DAUB16	3	5	90	90.00/100
COIF24	4	2	99	90.91/100
DAUB24	4	3	96	92.71/100
COIF6	4	4	93	92.47/100
DAUB22	4	5	90	92.22/100

Table 3.
Test results from AP2.

5.1 AP1

Thirty-five files from each class were collected and used. The tests were performed with those files not used for training, where two to five training files were experimented. For each set of training files, the DWT maximum level and mean decomposition level were tested with each of the 46 wavelet filters presented in **Table 2**. The best result, i.e., 96.88% accuracy, was obtained with Haar filter, showing excellent results and confirming our hypothesis of viability of using this system in coffee crop.

The cross-validation procedure was performed to present the best result in **Table 2**. The algorithm developed for cross-validation is illustrated in **Figure 8**.

5.2 AP2

To develop AP2, an application that uses digital image processing to estimate cicada density in a coffee crop was adopted, where 35 high-density class files, 35 low-density class files, and 35 files considered by an expert as zero density were used. In **Tables 3** and **4**, we present the results of the tests performed in the laboratory, which demonstrate the viability of future implantation in that system.

6. Conclusions

Both systems are being implemented in hardware for real-time coffee crop deployment using ESP8266 devices and their derivatives, integrating with cloud server for storing and organizing data to aid farmer decision-making. In future work, we must present practical results of their implementations.

Both AP1 and AP2 are systems that can be used as an additional form of coffee crop pest control and management. However, one possibility to be studied by the present research group is the combined use of the modalities in a single system in order to obtain even more improved results.

An important feature that was used for the laboratory experiments was the Java [18] Serialization class, which allowed for the wavelet transform in both AP1 and AP2 to be performed only once, storing its result on disk, making it possible to recover its value in repetitive cross-validation testing, substantially reducing equipment processing time.

	High density	**Low density**	**Null density**	**Total**
High density	30	3	0	33
Low density	1	31	1	33
Null density	0	0	33	33
Total	31	34	34	99

Table 4.
Best confusion matrix from AP2.

Author details

João Paulo Lemos Escola[1*†], Rodrigo Capobianco Guido[2†],
Alexandre Moraes Cardoso[1†], Douglas Henrique Bottura Maccagnan[3†],
João Marcelo Ribeiro[4†] and José Ricardo Ferreira Cardoso[1†]

1 Instituto Federal de São Paulo, Barretos, SP, Brazil

2 Instituto de Biociencias, Letras e Ciências Exatas, Univ. Estadual Paulista (UNESP) (São Paulo State University), SP, Brazil

3 Universidade Estadual de Goiás, Novo Horizonte II, Iporá, GO, Brazil

4 Instituto Federal do Sul de Minas Gerais, Muzambinho, MG, Brazil

*Address all correspondence to: jpescola@ifsp.edu.br

† These authors contributed equally.

References

[1] Mankin RW et al. Perspective and promise: A century of insect acoustic detection and monitoring. American Entomologist. 2011;**57**(1):30-44

[2] Martinelli NM, Zucchi RA. Cicadas (Hemiptera: Cicadidae: Tibicinidae) associated with coffee: Distribution, hosts and key to species (in Portuguese). Anais da Sociedade Entomolgica do Brasil. 1997:133-143

[3] De Souza JC. Coffee Cicada in Minas Gerais: Historical, Reconnaissance, Biology, Damage and Control (in Portuguese). Belo Horizonte: EPAMIG; 2007

[4] Maccagnan DHB. Cicada (Hemiptera: Cicadidae): Emergence, Acoustic Behavior and Sound Trap Development (in Portuguese) [Tese de Doutorado. PhD thesis]. Faculdade de Filosofia, Ciências e Letras da Universidade de São Paulo; 2008

[5] Haykin S. Neural Networks and Learning Machines. HAYKIN, Simon. 3/E. India: Pearson Education; 2010

[6] Guido RC. Effectively interpreting discrete wavelet transformed signals. IEEE Signal Processing Magazine. 2017; **34**(3):89-100

[7] Dawson DK, Efford MG. Bird population density estimated from acoustic signals. Journal of Applied Ecology. 2009;**46**(6):1201-1209

[8] Eliopoulos PA, Potamitis I, Kontodimas DC. Estimation of population density of stored grain pests via bioacoustic detection. Crop Protection. 2016;**85**:71-78

[9] Eliopoulos PA et al. Detection of adult beetles inside the stored wheat mass based on their acoustic emissions. Journal of Economic Entomology. 2015; **108**(6):2808-2814

[10] Marques TA et al. Estimating animal population density using passive acoustics. Biological Reviews. 2013; **88**(2):287-309

[11] Gardiner T, Hill J. A comparison of three sampling techniques used to estimate the population density and assemblage diversity of Orthoptera. Journal of Orthoptera Research. 2006: 45-51

[12] Langer F et al. Geometrical stem detection from image data for precision agriculture. 2018. arXiv preprint arXiv: 1812.05415

[13] Burgos-Artizzu XP et al. Real-time image processing for crop/weed discrimination in maize fields. Computers and Electronics in Agriculture. 2011;**75**(2):337-346

[14] Li Y et al. In-field cotton detection via region-based semantic image segmentation. Computers and Electronics in Agriculture. 2016;**127**: 475-486

[15] Burgos-Artizzu XP et al. Improving weed pressure assessment using digital images from an experience-based reasoning approach. Computers and Electronics in Agriculture. 2009;**65**(2): 176-185

[16] Bossi M, Goldberg E. Introduction to Digital Audio Coding and Standards. Springer Science & Business Media; 2012

[17] Guido RC. Paraconsistent feature engineering. IEEE Signal Processing Magazine. 2019;**36**(1):154-158

[18] Haverlock K. Object serialization, Java, and C++. Dr. Dobb's Journal: Software Tools for the Professional Programmer. 1998;**23**(8):32-35

Chapter 2

Artificial Intelligence Models to Predict the Influence of Linear and Cyclic Polyethers on the Electric Percolation of Microemulsions

Manuel Alonso-Ferrer, Gonzalo Astray Dopazo and Juan Carlos Mejuto

Abstract

This book chapter presents three predictive models, based on artificial neural networks, to determine the percolation temperature of different AOT microemulsions in the presence of different additives (crown ethers, glymes, and polyethylene glycols), which were developed in our laboratory by different authors. An artificial neural network model has been developed for each additive. The models developed, multilayer perceptron, were implemented with different input variables (chosen among the variables that define the packing or its chemical properties) and different intermediate layers. The best model for crown ethers has a topology of 10-8-1, for glymes the selected topology is 5-5-1, and for polyethylene glycol, the best topology was 5-8-8-5-1. The selected models are capable of predicting the electrical percolation temperature with good adjustments in terms of the root mean square error (RMSE), presenting values below 1°C for glymes and polyethylene glycols. According to these results, it can be concluded that the models presented good predictive capacity for percolation temperature. Nevertheless, the adjustments obtained for the crown ethers model indicate that it would be convenient to study new input variables, increase the number of cases, and even use other training algorithms and methods.

Keywords: microemulsion, electrical percolation, artificial neural, crown ethers, glymes, polyethylene glycols

1. Introduction

Microemulsions can be considered as the mixture of three, even four, basic components; two of these components have different polarity, while the third component, surfactant, gives the system thermodynamic stability (the presence of another surfactant, cosurfactant, facilitates the self-organization of the system is often needed) [1]. The authors affirm that as a result of this mixture, a microheterogeneous dynamic structure is obtained, with a continuous phase, constituted by the main solvent, in which microdroplets of a different phase are integrated (composed of the other solvent limited by surfactant molecules). By presenting two different hydrophobicity environments, microemulsions can be used as solvents for substances of different polarity [1].

Attending to Moldes et al. [1], the microemulsions show a low conductivity (10^{-9}–10^{-7} $\Omega^{-1}\cdot cm^{-1}$) [2, 3] and this remains stable while the temperature is below a specific value [1]. However, when the temperature reaches a certain value, the conductivity increases violently [1, 4, 5]. The point in which this event occurs is called percolation threshold [1]. This behavior is determined by the rigidity of the surfactant film that is determined by the way in which the surfactant molecules fit one with another and the presence of other substances, for example, additives [1, 6, 7]. It has been reported that the presence of additives in AOT microemulsions, for example, glymes and polyethylene glycols, facilitates the percolation and as a result reduces the percolation threshold [1, 2, 8], due to that additive are incorporated in the surfactant film and reduces their rigidity and the stabilizing forces [1].

This book chapter focuses on different artificial neural network (ANN) models developed in our laboratory to determine the percolation temperature of AOT/iC_8/H_2O microemulsions under the influence of three different additives, crown ethers [9] and linear polyethers (glymes and polyethylene glycols) [1].

ANNs are a modeling tool to capture complex relationships between input and output data [10]. ANNs are formed by interconnected artificial neuronal units [11] that simulate the biological neuron function, that is, use the input data in conjunction with a synaptic weight to generate an output response [12]. ANN models can provide good results in different fields such as: (i) in geology, to predict lithology in the subsurface [13]; (ii) in computing to schedule energy tasks in cloud data centers [14]; and (iii) in agricultural sciences to estimate Proctor parameters in soils [15] or in food technology to control the aging time in red wine [16], inter alia.

Nevertheless, the procedure to obtain a good prediction model is basically by trial and error, by varying the hidden layers' number and the input variables, the training cycles, among others [17]. According to this, ANN models require a lot of time analysis and computing power.

2. Material and methods

The crown ethers, glymes, and polyethylene glycols microemulsions databases used by Moldes et al. [1, 9] were in part compiled from the bibliography [2, 8, 18, 19]. The microemulsions have been prepared by weight [2, 8, 9, 18, 19] using reactants provided by Sigma-Aldrich [1, 2, 8, 9, 18, 19] and Fluka [1, 2, 8, 18, 19], and the percolation temperature has been determined by conductivity (with a Crison GPL conductivimeter) [2, 8, 18, 19].

All these microemulsions' composition remained constant with [AOT] = 0.5 $mol\cdot dm^{-3}$ (according to the total microemulsion volume) and W = [H_2O]/[AOT] = 22.2 [1, 2, 8, 9, 18, 19].

The crown ethers database (97 cases) was divided into two groups. The first group, training group (78 cases), was used to develop each prediction model, and a second group, validation group (19 cases), was used to choose the best model developed [9]. The glymes database (43 cases) was divided into the training group (32 cases) and the validation group (11 cases) [1]. Finally, for the polyethylene glycols database (82 cases), 68 cases were for the training group and 14 cases for the validation group [1].

2.1 Artificial neural networks

All models used a multilayer perceptron-type architecture composed of three different types of layers. The first layer is called input layer in which the data are received; the next layer/layers, called intermediate layer(s) (intended to process the information and transmit it to the following layers); and finally the output layer where all treated data are received and the predicted value is generated.

The general learning process of a neural network (shown below) is described in detail in the original papers [1, 9]. Data are presented to the neural network as a vector where x_n represents the input value in a neuron i Eq. (1).

$$x = (x_1, x_2, \ldots, x_n) \tag{1}$$

Then, the data are distributed to the first intermediate layer where, in each artificial neuron, two mathematical operations were carried out. These two processes are performed by the propagation function (Eq. (2)) and the activation function Eq. (3), where w_{ik} is the weight between the previous neuron i and the neuron k and $bias_k$ is the bias value for the neuron k. These two functions are responsible for receiving, unifying, processing, and sending information in the intermediate and output layers. All these processes can be observed in **Figure 1**.

$$S_k = \sum_{i=1}^{n} w_{ik} x_i + bias_k \tag{2}$$

$$y_k = \frac{1}{1 + e^{-S_k}} \tag{3}$$

During the training, the predicted value (obtained in the output neuron) is compared to the real value and the error is calculated.

This error is used in the next step to modify some parameters within the neural network and to try to improve the predicted value in the next neural cycle. This process is repeated according to the parameters established by the operator of the neural network and the training ends. Once the training is finished, the reservation data for validation are used to choose the best model developed according to the statistics used.

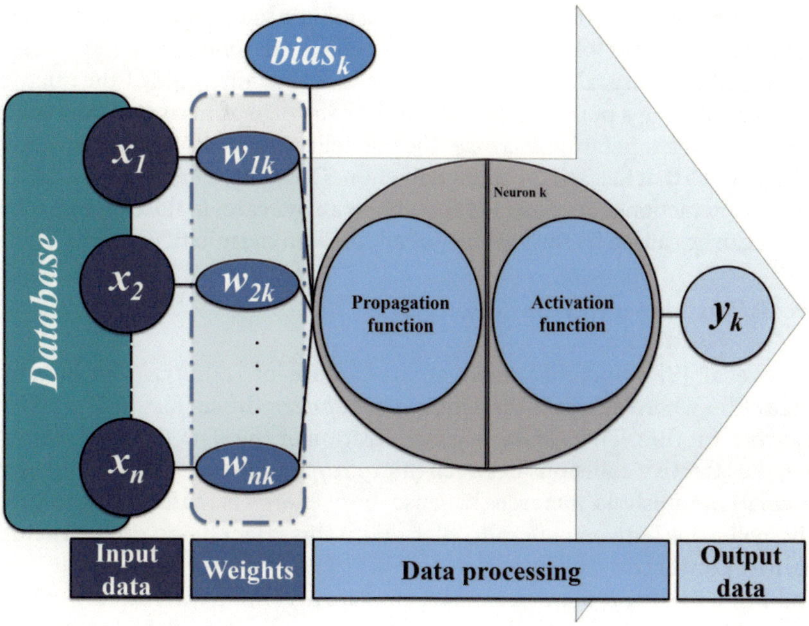

Figure 1.
Scheme of the operation for an artificial neuron.

Achieving a satisfactory model is a trial and error process in which the operator of the neural network must develop different models, for example, using different architectures (varying the number of neurons in the input layer and the intermediate layers). To establish a reasonable number of models, different approaches can be used to limit it. In the three models discussed here, the test range (δ) that determines the intermediate neurons number (Eq. (4)) has been used, where α corresponds to the number of training cases and β to the number of input variables [1, 9].

$$\frac{\alpha}{2\beta} < \delta < \frac{2\alpha}{\beta} \tag{4}$$

2.2 Statistical parameters and software use

Different statistical parameters can be used to check the model's adjustments. In our department, two of the most important are usually used: (i) the determination coefficient (R^2) (used to determine the correlation between the experimental and the predicted values) and (ii) the root mean square error (RMSE) (Eq. (5)),

$$RMSE = \sqrt{\frac{\sum_{i=1}^{n}\left(T_{ppred} - T_{pexp}\right)^2}{n}} \tag{5}$$

where T_p is the percolation temperature; the superscripts *pred* and *exp* correspond to the predicted value and the experimental value, respectively; and n is the number of cases [1, 9]. These parameters are used to choose the best model based on the values of the validation group.

All ANN models were developed with commercial software from Neural Planner Software [1, 9].

3. Results and discussion

According to Moldes et al. [1], temperature leads to electric percolation in microemulsions due to the transfer of charge and mass between collided microdroplets. This transfer is effective when microdroplets collide with enough energy (effective collisions) and a structural reorganization of the surfactant layers of the microdroplets occurs producing a pore through which the exchange of mass and load occurs [1].

An increase in temperature increases the possibility of effective collisions between microdroplets, so that facilitates the phenomenon. Certainly, a modification on the surfactant film interactions can cause a reduction, or an increase, in the effective collisions. This effect can be caused by the presence of additives in the microemulsion [1, 9].

3.1 Percolation temperature in crown ethers

Moldes et al. [9] report that crown ethers capture ions and transfer them to the surfactant film where decrease the repulsions between the surfactant's head groups in the surfactant film. This results in a stabilization of the structure and, consequently, the effective collisions between microdroplets decrease. For this reason, the percolation threshold increases because more energy is required to produce effective collisions between microdroplets, with this effect being predominant at low additive concentration [9].

In addition to this, Moldes et al. state that crown ethers can interact with the surfactant film causing destabilization, making the film less rigid, which facilitates effective collisions and favors the electric percolation [9]. This effect is given at high concentrations of additive [9, 18, 19].

3.2 Percolation temperature in glymes and polyethylene glycols

According to Dasilva-Carbalhal et al., glymes are considered as simply acyclic analogues from the group indicated above, the crown ethers, and can exhibit less cation affinity than their cyclic corresponding [2]. The glymes would be linked to the polar surfactant head group in AOT film, taking the place of water molecules in the hydration sphere [1, 2]. Due to the polar groups of the molecules would be located in the microdroplet aqueous core, the surfactant film reduces its rigidity, facilitating, and therefore, increasing the number of effective collisions and the percolation threshold will be achieved at lower temperatures [1].

Polyethylene glycols have a similar effect due to the molecule acting as a bridge between microdroplets and thus facilitating the percolation process [1, 2, 8, 20].

3.3 Artificial neural networks for crown ethers

To develop this neural model, Moldes et al. [9] selected different input variables to predict the percolation temperature, which was chosen for being related to the molecule's structure and its nature. The input variables were: (i) additive concentration -[Additive]-, (ii) log P, (iii) molecular mass -M_m–, (iv) maximum number of bonds between rings -B_M–, (v) minimum number of bonds between rings -B_m–, (vi) number of atoms that conforms a ring in a crown ether -A_{ring}–, (vii) number of heteroatoms -A_H–, (viii) number of oxygen atoms -nº O-, (ix) number of nitrogen atoms -nº N-, (x) number of benzene rings in the molecule -nº benz- [9].

The additive concentration supplies information about the influence of different quantities of additive [9]. Log P (partition coefficient between water and 1-octanol) provides information about the polarity (hydrophobicity) of a substance; finally, the other variables could provide information about molecular structure [9].

All these variables were also chosen, based on the knowledge obtained in previous works [21–23], and with the purpose to find which are the most important variables to determine the percolation temperature and develop a general model to be able to determine it with any type of additive [9].

As stated above, the trial and error method was used to determine the best artificial neural network. The best model developed by Moldes et al. [9], according to validation group, was the model with ten neurons in the first layer (input), eight neurons in the intermediate layer (hidden) and one neuron in the output layer (i.e., presents an architecture of 10-8-1) (**Figure 2**).

The selected ANN model presents, for the validation group, a determination coefficient around 0.724 [9]. Although this value is not very high, the root mean square error must be taken into account because it gives an idea of the deviation from the experimental value of the validation cases. In this sense, it can be seen that the RMSE is slightly greater than 1.1°C [9]. This value can be considered valid due to it being close to the limit of 1°C, which is, in our opinion, the barrier to the use of an ANN model. In **Figure 3**, the predicted values versus the experimental values of percolation temperature for validation group are shown (blue triangle). It can be seen, taking as reference the line with slope 1, that several points are far from the ideal prediction. It is these points that cause the RMSE in the validation group to exceed the threshold value of 1°C.

The ANN 10-8-1 presents, for the training group, a determination coefficient around 0.933 and RMSE of 1.625°C [9]. In **Figure 4**, the predicted values versus the experimental values of percolation temperature for training group are shown (blue triangle). It can be seen that different points are far from the ideal prediction, especially in the range 30–35°C.

The most important input variables to predict the percolation temperature of crown ethers are the additive concentration, the number of benzenes, and log P [9].

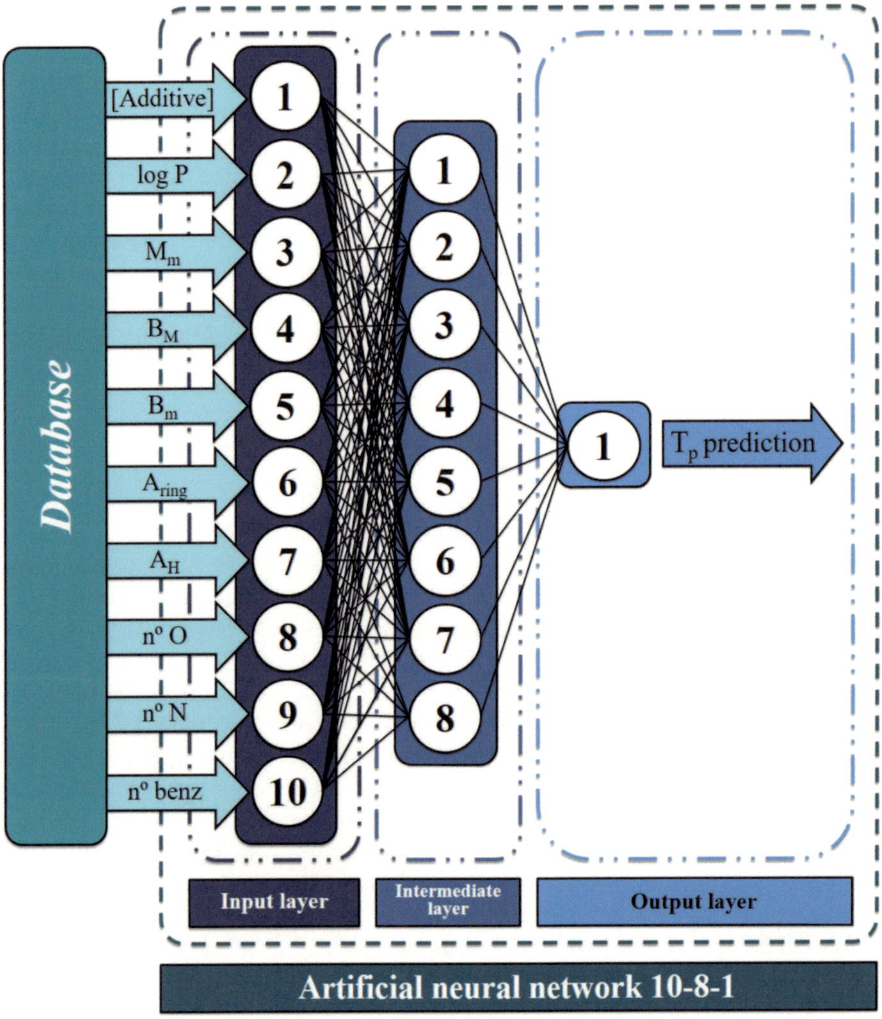

Figure 2.
Neural network (10-8-1) selected by Moldes et al. [9] to predict the percolation temperature for crown ethers.

According to the values obtained for the training group and the validation group, it can be said that the model developed could be used for the prediction of the percolation temperature. However, taking into account the RMSE values and the dispersion presented by some of the training and validation cases, it could be concluded that the model developed by Moldes et al. [9] should be improved to get the RMSE value to fall below 1, in both groups.

3.4 Artificial neural networks for glymes

Moldes et al. [1] selected for this ANN model different input variables based on the knowledge obtained in the previous works [9, 21–23]. In this case, the variables used by Moldes et al. [1] were: (i) additive concentration -[Additive]- and (ii) log P were used for being one of the most important variables in the previous model [9]. The variables (iii) molecular mass -M_m- and (iv) the number of oxygen in the molecule -nº O- were used because they had also been taken into account in the crown ethers model. Finally, (v) number of carbon atoms -nº C- (which had already been used in a previous model [23], and had shown good results) was taken into account.

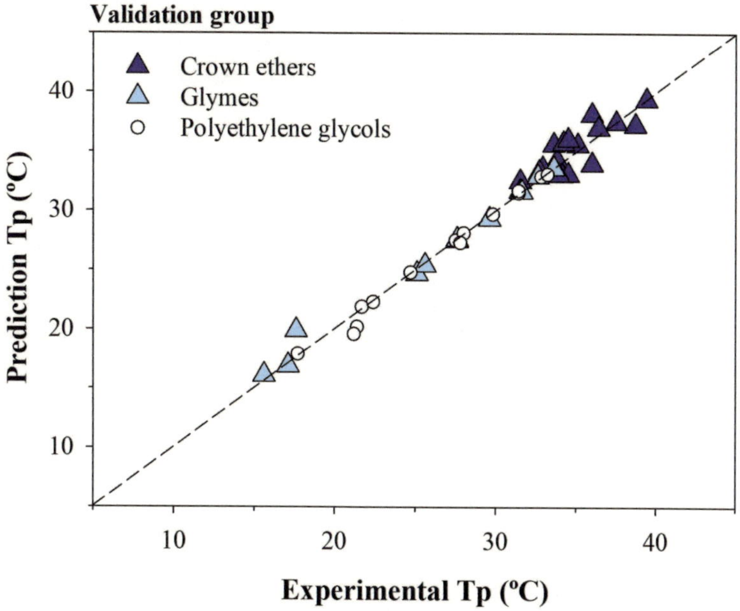

Figure 3.
Experimental and predicted values of percolation temperature for validation cases of crown ethers, glymes, and polyethylene glycols. The figure is a modification and combination of the data provided by Moldes et al. [1, 9]. Dashed line corresponds to the line with slope 1.

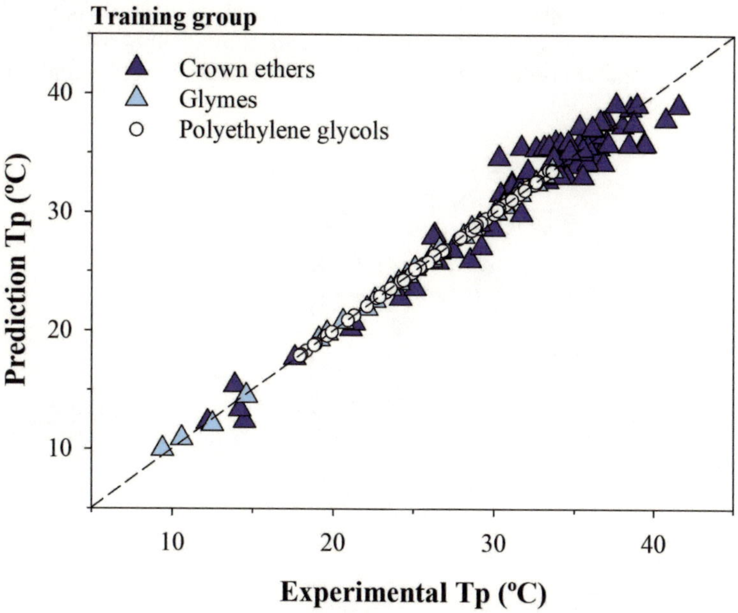

Figure 4.
Experimental and predicted values of percolation temperature for training cases of crown ethers, glymes, and polyethylene glycols. The figure is a modification and combination of the data provided by Moldes et al. [1, 9]. Dashed line corresponds to the line with slope 1.

Authors reported that the best model, taking into account the results of the validation group, is the model constituted by five, five and one neurons in the input, hidden and output layer, respectively (architecture 5-5-1, **Figure 5**) [1].

The selected model presents for validation group a determination coefficient with a value of close to 0.988 [1]. This high value corresponds to an RMSE value around 0.750°C [1] and it is considered a very good adjustment because it is below the reference error set at 1°C. For the training group, the selected neural network also has good adjustments, both in terms of R^2 (0.999) and the associated RMSE value (below 0.200°C) [1].

These good adjustments, both in determination coefficient and in the root mean square error, can be seen in **Figures 3** and **4,** where the predicted and experimental values for the validation and training groups are compared. It can be seen that for both the training and the validation, the points fit almost perfectly to the line with slope 1, except, perhaps, a point in the validation group (bottom left of graph 3) where a point that deviates slightly can be seen.

3.5 Artificial neural networks for polyethylene glycols

The last model proposed by Moldes et al. [1] is a neural network model with 5-8-8-5-1 architecture, that is, five input variables in the first layer; three hidden layers with 8, 8, and 5 neurons; and an output layer with one neuron, the percolation temperature (**Figure 6**).

The variables used were the same as those used in the model developed to predict the percolation temperature in glymes, that is: (i) additive concentration -[Additive]-, (ii) log P, (iii) molecular mass -M_m-, (iv) the number of oxygen in the molecule -nº O- and (v) the number of carbon atoms -nº C- [1].

This model provides a good determination coefficient (0.990) with an RMSE below 1°C for validation group [1]. This good fit for the validation group can also be observed for the training group (**Figures 3** and **4**). In these figures, it can be seen how data fit perfectly (R^2 of 0.999 [1]) to the slope line 1, with all points falling over the line. For the validation group, two points are observed outside but with a very small distance.

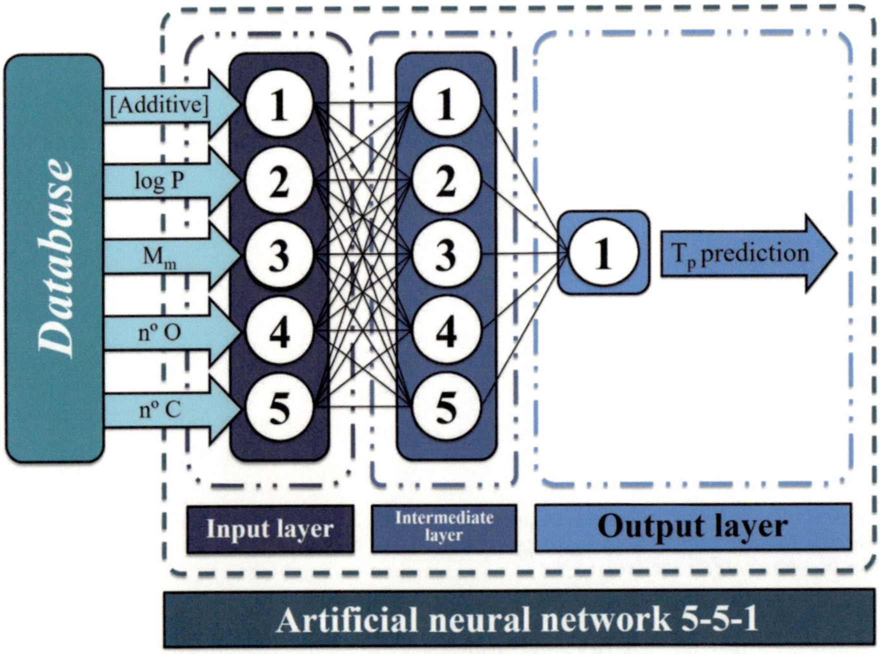

Figure 5.
Neural network (5-5-1) selected by Moldes et al. [1] to predict the percolation temperature for glymes.

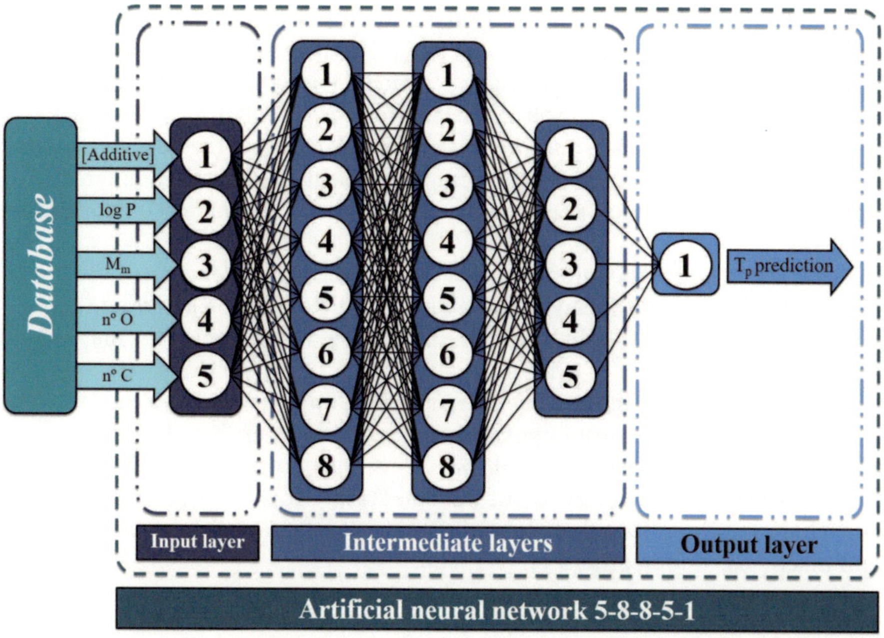

Figure 6.
Neural network (5-8-8-5-1) selected by Moldes et al. [1] to predict the percolation temperature for polyethylene glycols.

Once again, the RMSEs are below 1°C, which can be considered satisfactory, so that, the network could be used to determine the percolating temperature of polyethylene glycols.

Analyzing the results obtained by the authors, it can be verified that the most important input variable for the determination of the percolation temperature is the concentration of additive followed by log P. Attending to Moldes et al. [1], similar behavior is observed for both models regarding the importance of the variables. In these two models, the importance of the additive concentration clearly exceeds the importance of log P. All input variables remain in a range of similar importance for glymes and polyethylene glycols.

This behavior seems to indicate that the additive concentration and log P should be input variables to be taken into account whenever a percolation temperature prediction model is designed.

As stated earlier, the most important input variables for the crown ether model are additive concentration, benzene number, and log P; so, the idea of developing a single model to predict the percolation temperature of these types of substances should include necessarily these three variables.

Given the results obtained by the authors, it can be seen that the models developed to predict the temperature of percolation in glymes and polyethylene glycols present very good results; however, the predictability for the model developed for crown ethers offers more humble results.

4. Conclusions

The works collected in this book chapter are about prediction tools for percolation temperature under the presence of crown ethers, glymes, and polyethylene glycols as additives.

The predictive models were developed based on artificial neural networks and according to the data provided by Moldes al. [1, 9]: (i) the best model to predict the percolation temperature in crown ethers is a model with an architecture of 10-8-1 that presents for validation dataset an RMSE around 1.1°C, (ii) for glymes the architecture selected is 5-5-1, and (iii) for polyethyleneglycol the best architecture was 5-8-8-5-1. The last two models present a root mean square error under 1°C, which demonstrates its good performance and improves the model developed for crown ethers.

Taking into account the importance of the variables used for each of the models developed by the authors, it can be concluded that the inclusion of the additive concentration and log P should be required for any general model that may arise in the future.

To finish, it would be interesting to try the development of a joint model using the variables with the greatest importance of the models developed in our laboratory; it also would be very convenient to improve the new model with the inclusion of new experimental cases, new input variables, other training algorithms, and even including different machine learning models such as random forest, support vector machines, among others.

Acknowledgements

M. Alonso-Ferrer thanks the University of Vigo for his contract supported by FEADER 2018/002B (Xunta de Galicia, Consellería de Medio Rural, project "Desarrollo de modelos de predicción de origen en vinos de denominaciones de origen gallegas"). Gonzalo Astray thanks the University of Vigo for his contract supported by "Programa de retención de talento investigador da Universidade de Vigo para o 2018" budget application 0000 131H TAL 641.

Author details

Manuel Alonso-Ferrer, Gonzalo Astray Dopazo* and Juan Carlos Mejuto
Department of Physical Chemistry, Faculty of Sciences, University of Vigo, Ourense, Spain

*Address all correspondence to: gastray@uvigo.es

References

[1] Moldes ÓA, Cid A, Montoya IA, Mejuto JC. Linear polyethers as additives for AOT-based microemulsions: Prediction of percolation temperature changes using artificial neural networks. Tenside, Surfactants, Detergents. 2015;**52**(4):264-270

[2] Dasilva-Carbalhal J, García-Río L, Gómez-Díaz D, Mejuto JC, Pérez-Lorenzo M. Influence of glymes upon percolative phenomena in AOT-based microemulsions. Journal of Colloid and Interface Science. 2005;**292**(2):591-594

[3] Eastoe J, Robinson BH, Steytler DC, Thorn-Leeson D. Structural studies of microemulsions stabilised by aerosol-OT. Advances in Colloid and Interface Science. 1991;**36**:1-31

[4] Eicke H-F, Borkovec M, Das-Gupta B. Conductivity of water-in-oil microemulsions: A quantitative charge fluctuation model. The Journal of Physical Chemistry. 1989;**93**(1):314-317

[5] Feldman Y, Kozlovich N, Nir I, Garti N. Dielectric relaxation in sodium bis(2-ethylhexyl)sulfosuccinate-water-decane microemulsions near the percolation temperature threshold. Physical Review E. 1995;**51**(1):478-491

[6] García-Río L, Leis R, Mejuto JC, Peña ME, Iglesias E. Effects of additives on the internal dynamics and properties of water/AOT/isooctane microemulsions. Langmuir. 1994;**10**(6):1676-1683

[7] García-Río L, Hervés P, Mejuto JC, Pérez-Juste J, Rodríguez-Dafonte P. Effects of alkylamines on the percolation phenomena in water/AOT/isooctane microemulsions. Journal of Colloid and Interface Science. 2000;**225**(2):259-264

[8] Arias-Barros SI, Cid A, García-Río L, Mejuto JC, Morales J. Influence of polyethylene glycols on percolative phenomena in AOT microemulsions. Colloid & Polymer Science. 2010;**228**(2):217-221

[9] Moldes ÓA, Cid A, Astray G, Mejuto JC. Percolative behavior models based on artificial neural networks for electrical percolation of AOT microemulsions in the presence of crown ethers as additives. Tenside, Surfactants, Detergents. 2014;**51**(6):533-540

[10] Azizi A, Abbaspour-Gilandeh Y, Nooshyar M, Afkari-Sayah A. Identifying potato varieties using machine vision and artificial neural networks. International Journal of Food Properties. 2016;**19**(3):618-635

[11] Dawson CW, Wilby RL. Hydrological modelling using artificial neural networks. Progress in Physical Geography. 2001;**25**(1):80-108

[12] Wu H, Tian L, Chen B, Jin B, Tian B, Xie L, et al. Verification of imported red wine origin into China using multi isotope and elemental analyses. Food Chemistry. 2019;**301**:125137

[13] Kamenski A, Cvetković M, Kolenković Močilac I, Saftić B. Lithology prediction in the subsurface by artificial neural networks on well and 3D seismic data in clastic sediments: A stochastic approach to a deterministic method. GEM - International Journal on Geomathematics. 2020;**11**(1):8

[14] Sharma M, Garg R. An artificial neural network based approach for energy efficient task scheduling in cloud data centers. Sustain Comput Informatics Syst. 2020;**26**:100373

[15] Bayat H, Asghari S, Rastgou M, Sheykhzadeh GR. Estimating proctor parameters in agricultural soils in the Ardabil plain of Iran using support

vector machines, artificial neural networks and regression methods. Catena. 2020;**189**:104467

[16] Astray G, Mejuto JC, Martínez-Martínez V, Nevares I, Alamo-Sanza M, Simal-Gandara J. Prediction models to control aging time in red wine. Molecules. 2019;**24**(5):826

[17] Iglesias-Otero MA, Fernández-González M, Rodríguez-Caride D, Astray G, Mejuto JC, Rodríguez-Rajo FJ. A model to forecast the risk periods of Plantago pollen allergy by using the ANN methodology. Aerobiologia (Bologna). 2015;**31**(2):201-211

[18] Dasilva-Carvalhal J, Fernández-Gándara D, García-Río L, Mejuto JC. Influence of aza crown ethers on the electric percolation of AOT/isooctane/water (w/o) microemulsions. Journal of Colloid and Interface Science. 2006;**301**(2):637-643

[19] Dasilva-Carvalhal J, García-Río L, Gómez-Díaz D, Mejuto JC, Rodríguez-Dafonte P. Influence of crown ethers on the electric percolation of AOT/isooctane/water (w/o) microemulsions. Langmuir. 2003;**19**(15):5975-5983

[20] Mehta SK, Sharma S. Temperature-induced percolation behavior of AOT reverse micelles affected by poly(ethylene glycol)s. Journal of Colloid and Interface Science. 2006;**296**(2):690-699

[21] Cid A, Astray G, Manso JA, Mejuto JC, Moldes OA. Artificial intelligence for electrical percolation of aot-based microemulsions prediction. Tenside, Surfactants, Detergents. 2011;**48**(6):477-483

[22] Montoya LA, Astray G, Cid A, Manso JA, Moldes OA, Mejuto JC. Influence prediction of small organic molecules (Ureas and Thioureas) upon electrical percolation of AOT-based microemulsions using artificial neural networks. Tenside, Surfactants, Detergents. 2012;**49**(4):316-320

[23] Moldes ÓA, Astray G, Cid A, Iglesias-Otero MÁ, Morales J, Mejuto JC. Percolation threshold of AOT microemulsions with n-alkyl acids as additives prediction by means of artificial neural networks. Tenside, Surfactants, Detergents. 2013;**50**(5):360-368

Chapter 3

Increasing the Efficiency of Rule-Based Expert Systems Applied on Heterogeneous Data Sources

Juan Ignacio Guerrero Alonso, Enrique Personal, Antonio Parejo, S. García, Antonio Martín and Carlos León

Abstract

Nowadays, the proliferation of heterogeneous data sources provided by different research and innovation projects and initiatives is proliferating more and more and presents huge opportunities. These developments create an increase in the number of different data sources, which could be involved in the process of decision-making for a specific purpose, but this huge heterogeneity makes this task difficult. Traditionally, the expert systems try to integrate all information into a main database, but, sometimes, this information is not easily available, or its integration with other databases is very problematic. In this case, it is essential to establish procedures that make a metadata distributed integration for them. This process provides a "mapping" of available information, but it is only at logic level. Thus, on a physical level, the data is still distributed into several resources. In this sense, this chapter proposes a distributed rule engine extension (DREE) based on edge computing that makes an integration of metadata provided by different heterogeneous data sources, applying then a mathematical decomposition over the antecedent of rules. The use of the proposed rule engine increases the efficiency and the capability of rule-based expert systems, providing the possibility of applying these rules over distributed and heterogeneous data sources, increasing the size of data sets that could be involved in the decision-making process.

Keywords: rule-based expert system, inference engine, heterogeneous data source integration, distributed data sources

1. Introduction

The expert systems (ESs) are one of the most traditional artificial intelligence techniques [1], providing the possibility of implementing systems which allow us to solve problems in a limited domain. However, the variety and possibilities of ESs have improved in the last years with the combination of them with other technologies such as fuzzy logic [2], Bayesian network [3], etc. Moreover, several languages and tools have been developed to provide faster developments (e.g., CLIPS, LISP, PROLOG, etc.) and deployments. Other systems are improved in collaboration with other technologies in order to expand the domain of problems and increase the knowledge base. As example, [4] improves the capabilities of an ES adding text mining, neural networks, and statistical techniques.

On the one hand, the scope of ES application is very extensive, including health [5], education [6], physics [7], chemical [8], mechanics [9], etc. Thus, the application of ES is not only restricted to solve the problem, they usually include an explaining engine, which could be used for educational purposes, operating at the same time.

The ESs have a limited domain, and the size of used data sets is smaller than in other artificial intelligence techniques. However, the improvement of availability of information, new concepts related to sensor networks, and the capability to generate and consume information in different sectors provide a different scenario, in which the ESs traditionally had a lot of information disseminated into different information resources. Although each information resources could have its own structure, the stored information could be very similar. In this scenario, it is essential to make the analysis of the distributed data sets and apply the different rules and inference engines in these distributed data sets possible.

In response to this problem, an additional layer is proposed in the current chapter which could be added to the ES engines, mainly based on rules and fuzzy logic. This new layer or middleware allows the ES to make a metadata integration of heterogeneous data sets, without making a real integration and replication of the information, allowing by mean edge, and computing the application of rules in the disseminated data sets by means of a logical decomposition of rules based on metadata integration results. This novel layer or middleware, named distributed rule engine extension (DREE) has an application protocol interface (API) to allow the communication between the ES engine and the different data sources. The DREE enables the ES engines to be applied in heterogeneous data sets disseminated in a network, by means of installation of edge computing daemon (ECD) in each distributed node.

Edge computing has some similarities with fog computing, cloud computing, etc. [10] provides a review of different similar technologies, providing a very complete survey. In the case of the proposed system, edge computing is usually related to the Internet of Things (IoT) devices. In case of the proposed solution, the edge nodes are the computers or devices, which has the information stored, by means of any type of database management system.

In the following sections, the general architecture is detailed. Firstly, the process of metadata integration from heterogeneous data sets is described. Secondly, the process of logical decomposition and how the edge computing contributes in the process of distributed application of antecedent rule are described. Finally, the experimental application of the proposed layer or middleware is shown, with the conclusion of results and future research lines.

2. Architecture overview

The application process of the DREE is performed in three stages. In the first one, the metadata integration is performed. In the second, the rules from knowledge base are logically decomposed, according to the metadata model. In the third one, the rules are carried out in the disseminated data sets by means of edge computing in the distributed nodes. In this sense, to carry out these tasks, the architecture proposed for the DREE is shown in **Figure 1**.

Each node has an edge computing daemon which performs queries in the database node to gather all information from local data sets, sending only metadata to the DREE. This information is integrated and translated in a unified metadata structure to an engine directly available for DREE.

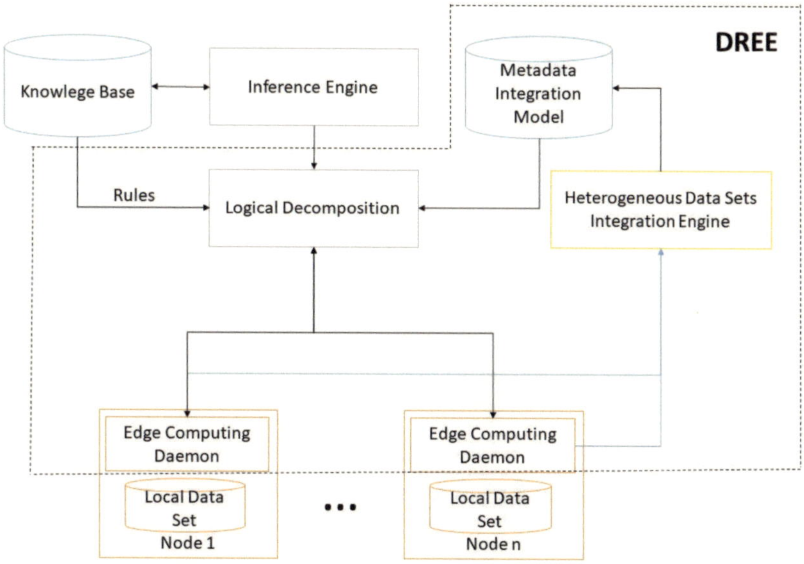

Figure 1.
DREE architecture and flow overview.

The metadata integration model database provides a map about the distribution of data between the different nodes. This information is used by logical decomposition engine (LDE) to identify the dependences between variables and the complexity of rules in the knowledge base. In some cases in which there exists a very complex logic expression with a high level of dissemination between nodes, the information would be stored in this database in order to make the rule application, but this option needs to be configured manually.

3. Heterogeneous data set integration (HDSI)

The fundamentals of heterogeneous data set integration were described in [11]. In this reference, a heterogeneous data source integration system (HDSIS) is described and applied to smart grid and health. Following this idea, a HDSIS evolution is implemented providing also a logic integration of all information at metadata level. The architecture of the proposed HDSI is shown in **Figure 2**. Specifically, the metadata from different sources is the only information that is integrated and stored.

The HDSI includes a metadata mining engine (MME), which connects with EDCs by means of characterization engine, in order to extract information from local databases in each node, getting new and integrating existing useful metadata. The dynamic extract, transform, and load (ETL) engine performs the process to integrate the metadata, previously inferred by the MME, according to the specifications gathered by metadata mining engine and the rules stored in the knowledge base. All these modules define the query engine and the rule-based expert system. The query engine is focused on performing the different tasks required to the queries in the distributed resources. The rule-based expert system (RBES) is included in the HDSI and implements the rules that perform an integration of all metadata from all disseminated resources. Therefore, it is an RBES oriented to information integration. Some references show the application of HDSIS in different problems: nontechnical losses detection [12], electric vehicle and consumption modeling in smart grids [13], etc.

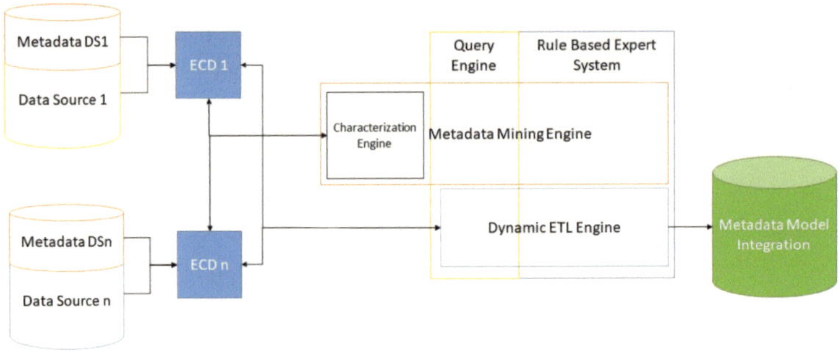

Figure 2.
Heterogeneous data set integration module overview.

The ECD includes some modules to perform the necessary queries over the database. Thus, the architecture of ECD is described in the next section.

4. Logical decomposition engine

The logical decomposition engine included two parts. The main part is in the LDE and is shown in **Figure 3**. The second part is located in the ECDs, its structure being shown in **Figure 4**.

The LDE has a first component, named logic parser. This component makes it possible to parse logic expression, which could be based on fuzzy logic or type-2 fuzzy logic, too. The logic parser works with different mathematical representation standard languages: MathML (mathematical markup language v3.0 [14]) and the OpenMath standard [15]. The logic parser serves as a RESTful web service interface, which supports extensible markup language (XML) format messages based on MathML and JavaScript Object Notation (JSON) format messages based on OpenMath.

The LDE has registered some logic expressions, which are typical equations and other previously performed expressions. LDE database stores all information about the decomposition of logic expression. In this sense, if the expression is not yet in the database, the decomposition engine checks the dependency between variables and the number of steps necessary to calculate the result. To reduce the number of steps, the decomposition engine applies a particle swarm optimization (PSO) [16] algorithm. The objective of using a genetic algorithm is to find out an equivalent logic expression with a small number of steps, reducing the dependency between variables and operations.

Thus, if the number of steps is still high or the dependence between variables could involve different data sets in disseminated nodes, the subscheduler plans the message exchange in order to establish the message sequence. This message sequence could be provided by the result of partial or complete logic expression. Thus, it is possible that some logic expressions take a long time to get the result, because of the high number of messages exchanged. In this case, the user could manually configure the HDSI to integrate the anonymized information in a server in order to reduce the time and consumption of edge computing nodes. This configuration and specification about decomposition are stored in expression database. When the LDE detects an expression of this type, the scheduler manages the decomposition according to the manual configuration.

In the ECD, the messages from the LDE are interpreted and transferred to local SQL engine and logic inference engine. The local database engine extracts the required information from local database and, this information is gathered by

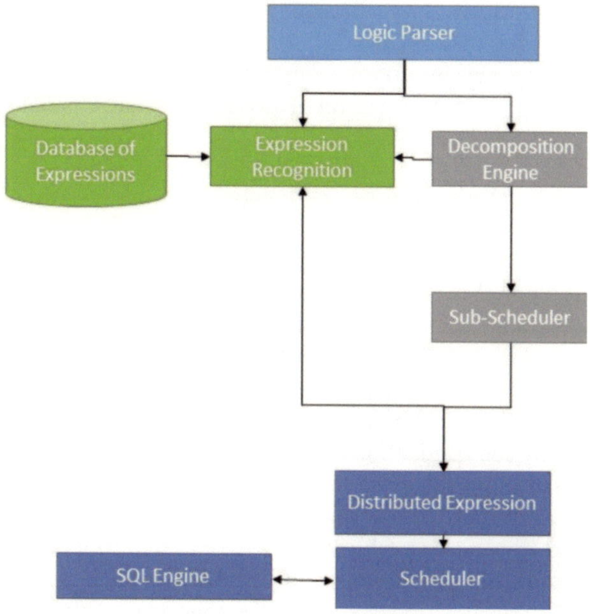

Figure 3.
The main part of LDE.

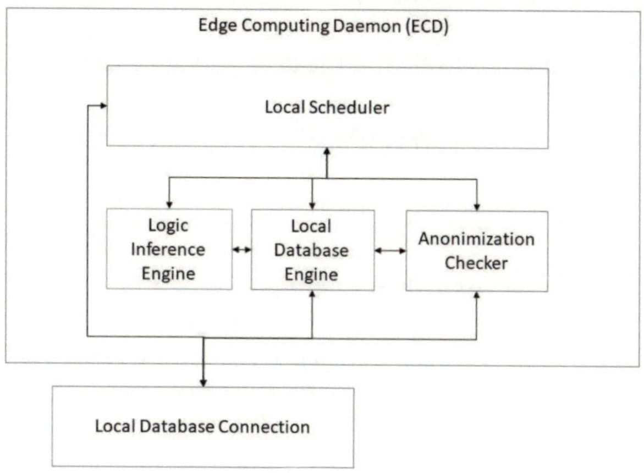

Figure 4.
Functional architecture of the edge computing daemon.

the logic inference engine which performs the expressions and returns the results. Simultaneously, the extracted information and results are checked by the anonymization checker, which is responsible for tests if the information contains data, which would be anonymized.

The local database engine is formed by two mechanisms. One engine is based on simple standard statement query language (SQL). There are several database management systems that include improvements in the SQL language. The proposed engine only applies queries on standard format, using minimized SQL queries with simple statements to ensure the compatibility. The other engine is based on NoSQL language, and it was designed to operate with unstructured databases.

Additionally, the local database engine notifies any modification in the data set to the LDE, in order to update the metadata model. In this case, the LDE recalculates the decomposition of different expressions that involves the data from the updated or modified data set.

The logic inference engine supports logic, fuzzy logic, and type-2 fuzzy logic expressions. This engine does not perform any decompositions but only performs the expressions provided by LDE.

4.1 Particle swarm optimization

The prioritization algorithm works as a swarm intelligence algorithm. The application of the algorithm is performed after a preprocessing of information.

The prioritization algorithm is based on the parametric optimization until a solution is obtained. This optimization is executed depending on the capabilities of a system. The canonical PSO model consists of a swarm of particles, which are initialized with a population of random candidate solutions. The candidate solutions are generated by the application of different properties oriented to reduce the dependence and operations involved in the logic expression to lead to the expression, which minimizes the number of messages and distributed expression. They iteratively move through the d-dimension problem space to search for the new solutions, where fitness f can be calculated as the certain quality measure. Each particle has a position that is represented by the position-vector x_{id} (i is the index of the particle, and d is the dimension) and a velocity represented by the velocity-vector v_{id}. Each particle remembers its best position in the vector $x_{i\#}$, and its j-th dimensional value is $x_{\#ij}$. The best position-vector among the swarm is stored in the vector x_{*}, and its j-th dimensional value is x_{*j}. At the iteration time t, the update of the velocity from the previous velocity to the new velocity is determined by Eq. (1). The new position is determined by the sum of the previous position, and the new velocity is determined by Eq. (2):

$$v_{id}(t+1) = w \cdot v_{id}(t) + c_1 \cdot \psi_1 \cdot (p_{id}(t) - x_i(t)) + c_2 \cdot \psi_2 \cdot (p_g(t) - x_{id}(t)) \quad (1)$$

$$x_{id}(t+1) = x_{id}(t) + v_{id}(t+1) \quad (2)$$

where c_1 and c_2 are constant weight factors, p_i is the best position achieved by particle i, p_g is the best position obtained by the neighbors of particle i, ψ_1 and ψ_2 are random factors in the [0,1] interval, and ω is the inertia weight. Some references denote c_1 and c_2 as the self-recognition component and the coefficient of the social component, respectively.

Different constraints can be applied to ensure the convergence of the algorithm. In this case, the operations are oriented to optimize the fitness, and the fitness is calculated based on the number of distributed operations and number of exchanged messages.

PSO algorithm:

1. Initialize particles

2. Repeat

   a. Calculate the fitness values for each particle.

   b. Is the current fitness value better than p_i?

      i. Yes. Assign the current fitness as the new p_i.

      ii. No. Keep the previous p_i.

c. Assign the best particle's p_i value to p_g.

d. Calculate the velocity for each particle.

e. Use each particle's velocity value to update its data values.

3. Until stopping criteria are satisfied

5. Testing DREE

The proposed system DREE was tested with different RBESs. These tests were performed inserting the DREE between RBES and data sets, redefining the inference engines to replace the calls to rule execution by the DREE interface. [17] provides the description of different ESs oriented to health, telecommunication, power supply, etc. The proposed architecture is shown in **Figure 5**. The information used in the different ESs was manually disseminated among five different nodes. The DREE was able to work with different ESs at the same time. The application of all rules from all ESs at the same time may involve information from different domain knowledge, because the DREE only applies the rules that the corresponding inference system provides. Thus, the state and the decision supported is performed inside of the ES. The DREE only returns the result of consequent from the fired rule. Additionally, the usage of ECD makes independent from the local data set management systems. Therefore, in these cases, the ESs improved their operational capabilities, providing the possibility to operate in real time. Additionally, the ESs may simultaneously run their own rule-based knowledge base, increasing the reliability of the different systems.

The usage of the PSO algorithm instead using the most simplified logic expression shows an increase of efficiency of the distributed operations, reducing the operations and message exchanging in 20% related to initial fuzzy and type-2 fuzzy logic expressions and in 5% related to initial logic expressions. This fact is due to how the data is disseminated by the different nodes; different distributions of data between nodes provide different optimizations. Thus, if the number of nodes or

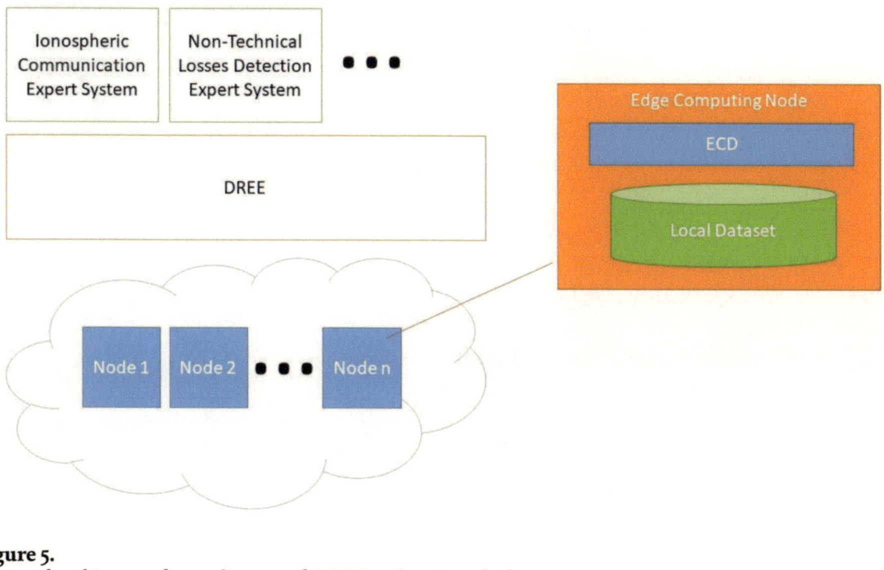

Figure 5.
Proposed architecture for application of DREE to the ESs in [17].

data sets from the nodes is updated or modified, it is necessary to recalculate the decomposition of logic expression.

In other cases, like [4, 18], the integration of the proposed DREE decreased the hardware requirements related to the storing systems.

Although the system increases the message exchange, the system avoids integrating all information in a centered data base, without using big data infrastructure, taking advantage from edge computing infrastructure and distributed nodes. Additionally, the system can quickly react to any updating or modification in any data set from the nodes involves in the DREE.

6. Conclusions

The proposed DREE provides the opportunity to integrate a great quantity of information in the inference engine, without the requirement of a big data infrastructure and the extract, transform, and load to physically integrate all the data sets. DREE makes a metadata-level integration. At this level the integration is quicker and smaller, and it does not need a great quantity of hard disk or memory space. Although the message exchanging increases the volume of the exchanged information, the load of edge computing nodes is optimized in the LDE, before the application of rules.

The deployment of DREE is simplified by adding a RESTful web service interface to access and replace the traditional services consumed by the inference engine. Additionally, the deployment on the edge nodes is summarized to install a daemon, named ECD, which simplifies the access to information in the local nodes, notifying any modification or updating in the different data sets, by updating the metadata model in the LDE. Thus, DREE reacts to any change in the data sets located on edge nodes.

Finally, the proposed DREE is independent from the ES, providing the possibility to run simultaneously several ESs. Thus, the processing load balancing is automatically provided by the information dissemination around the nodes.

Author details

Juan Ignacio Guerrero Alonso*, Enrique Personal, Antonio Parejo, S. García, Antonio Martín and Carlos León
Department of Electronic Technology, University of Seville, Seville, Spain

*Address all correspondence to: juaguealo@us.es

References

[1] Jabbar HK, Khan RZ. Survey on development of expert system in the areas of medical, education, automobile and agriculture. In: 2015 2nd International Conference on Computing for Sustainable Global Development (INDIACom); 2015. pp. 776-780

[2] D'Aquila RO, Crespo C, Mate JL, Pazos J. An inference engine based on fuzzy logic for uncertain and imprecise expert reasoning. Fuzzy Sets and Systems. 2002;**129**(2):187-202

[3] Chojnacki E, Plumecocq W, Audouin L. An expert system based on a Bayesian network for fire safety analysis in nuclear area. Fire Safety Journal. 2019;**105**:28-40

[4] Guerrero JI, León C, Monedero I, Biscarri F, Biscarri J. Improving knowledge-based systems with statistical techniques, text mining, and neural networks for non-technical loss detection. Knowledge-Based Systems. 2014;**71**:376-388

[5] Jimenez ML, Santamaría JM, Barchino R, Laita L, Laita LM, González LA, et al. Knowledge representation for diagnosis of care problems through an expert system: Model of the auto-care deficit situations. Expert Systems with Applications. 2008;**34**(4):2847-2857

[6] Nagata T, Sasaki H. Personal computer based expert system for power system operation education. International Journal of Electrical Power & Energy Systems. 1996;**18**(3):195-201

[7] Végh J. A simple "embedded" reasoning inference engine with application example in the X-ray photoelectron spectroscopy. Computer Physics Communications. 2004;**160**(1):8-22

[8] Qian Y, Li X, Jiang Y, Wen Y. An expert system for real-time fault diagnosis of complex chemical processes. Expert Systems with Applications. 2003;**24**(4):425-432

[9] Magalhães SC, Borges RFO, Calçada LA, Scheid CM, Folsta M, Waldmann A, et al. Development of an expert system to remotely build and control drilling fluids. Journal of Petroleum Science and Engineering. 2019;**181**:106033

[10] Yousefpour A, Fung C, Nguyen T, Kadiyala K, Jalali F, Niakanlahiji A, et al. All one needs to know about fog computing and related edge computing paradigms: A complete survey. Journal of Systems Architecture. 2019;**98**:289-330

[11] Guerrero JI, García A, Personal E, Luque J, León C. Heterogeneous data source integration for smart grid ecosystems based on metadata mining. Expert Systems with Applications. 2017;**79**:254-268

[12] Guerrero JI, Personal E, Parejo A, Monedero I, Biscarri F, Biscarri J, et al. High performance data analysis for non-technical losses reduction. En: Lou J, editor. Smart Grids: Emerging Technologies, Challenges and Future Directions. New York, USA: Nova Science Publishers; 2017. p. 1-45. (Energy Science, Engineering and Technology)

[13] Guerrero JI, García A, Personal E, Parejo A, Pérez F, León C. A Rule-based expert system for heterogeneous data source integration in smart grid systems. En: Ryan D, editor. Expert Systems: Design, Applications and Technology. New York, USA: Nova Science Publishers; 2017. p. 59-104. (Computer Science, Technology and Applications)

[14] Mathematical Markup Language (MathML) Version 3.0 2nd Edition

[Internet]. [cited 10th November 2019]. 2014. Available in: https://www.w3.org/TR/MathML3/

[15] The OpenMath Standard [Internet]. [cited 10th November 2019]. Available in: https://www.openmath.org/standard/om20-2019-07-01/omstd20.html

[16] Guerrero JI, Personal E, García A, Parejo A, Pérez F, León C. Distributed charging prioritization methodology based on evolutionary computation and virtual power plants to integrate electric vehicle fleets on smart grids. Energies. 2019;**12**(12):2402

[17] Monedero I, Martín A, Elena J, Guerrero J, Biscarri F, León C. A practical overview of expert systems in telecommunication networks, medicine and power supplies. In: Expert System Software: Engineering, Advantages and Applications. New York: Nova Science Publishers; 2012. p. 178-210

[18] Guerrero J, Parejo Matos A, Personal E, Monedero I, Biscarri F, Biscarri J, et al. From rule based expert system to high-performance data analysis for reduction of non-technical losses on power grids. International Journal on Advances in Intelligent Systems. 2017;**10**:136-146

Chapter 4

A Conceptual Framework for Modeling Smart Parking

Brahim Lejdel

Abstract

Vehicles are highly used in the city. If the drivers of vehicles have an appointment in the city, they are looking for parking. Thus, they need to know where to find one in real time. In this paper, we present a smart model that is based on a combination of multi-agent system and genetic algorithm (MAS-GA). The smart model can help the drivers find the optimal parking when the drivers make a request for parking according to their position on the road and the waiting and parking time. This smart model is based on four parameters: the availability of parking, cost of parking, the distance between the actual position of the vehicle and the destination parking, and traffic congestion. We can also add the time to arrive a destination parking. Thus, the proposed smart model helps to maximize the utilization of space resources of a city as parking and reduce the waiting and parking time.

Keywords: vehicle agent, parking agent, MAS-GA approach, smart model, smart city

1. Introduction

Parking in the city can be an attractive subject because 30% of traffic congestion is caused by drivers circling to find an available parking for their vehicles [1]. Thus, the vehicle needs to have a decision system that allows it to find the optimal parking in a short time. Economic income is also affected by this decision system that can manage and control the availability of parking in real time; in the same time, they often lose revenue due to inadequate meter enforcement and no parking, standing, and loading zone violations [2]. We will treat two main questions in this paper. The first is "what is the optimal method to manage, reduce, and control the parking in the city?" The second is "what is the optimal strategy which permits finding the optimal parking?"

In this paper, we propose to use a multi-agent system which allows distributing the different tasks between the agents when each agent can perform genetic algorithms to find the optimal parking in real time, thus adapting rapidly to traffic congestion the availability of place in the parking. Thus, all agents can cooperate and negotiate to find the best solution which can plan for allocation of vehicle in parking. Then, we develop a GIS system which allows knowing the position of the vehicle and all data associated with it as vehicle id, waiting time, and distance between proposed parking and vehicle.

After a deep study of the subject, we have found three factors that can affect the allocation of vehicles into the optimal parking, as the waiting and parking time, parking costs, distance between the vehicle and proposed parking.

IntechOpen

This paper is organized as the following. Firstly, we will present a state-of-the-art review for allocation of vehicle into parking. Then, we describe our proposed approach which is based on two approaches, the multi-agent system and genetic algorithm (MAS-GA). Finally, we add some experimental results and conclusion.

2. Related works

Many works have been studied to propose an intelligent parking. In this section, we will introduce the most important works that treat this issue.

Stéphane et al. [3] propose in this paper a new system that integrates the IoT and a predictive model based on ensemble methods to optimize the prediction of the availability of parking spaces in smart parking. The tests that we carried out on the Birmingham parking data set allowed reaching a mean absolute error (MAE) of 0.06% on average with the algorithm of bagging regression (BR). These results have improved the best existing performance by over 6.6% while dramatically reducing system complexity.

Diya and Binsu [4] propose a new prototype for the smart vehicle parking system. A genetic algorithm approach has been taken to address the issue of scheduling the vehicle to the parking bay.

Li et al. [5] propose to use modern intelligent agents having the capability of planning, mobility, execution monitoring, and coordination, which take into account the negotiable space on parking prices, make a strategic decision by adopting the intelligent agent system, and then select the optimal parking for the driver. These characteristics can be utilized to build an integrated parking assistant system. The autonomous coordination activities challenge traditional approaches and call for new paradigms and supporting middleware. This agent approach is based on the coordination network. This coordination network is truly bringing benefit to drivers and parking operators.

Oyentaryo and Pasquier [6] proposed an approach that is to design a self-training system that makes use of human expertise to automatically derive a working vehicle control system. A new neuro-fuzzy architecture known as the GenSoYager fuzzy neural network has been realized and integrated with a vehicle driving simulator for training and testing purposes. This approach is based on detecting system, motion planning, and supplying information to detect the best parking.

Tang et al. [7] describe a wireless sensor network (WSN)-based intelligent car parking system. In the system, low-cost wireless sensors are deployed in a car park field, with each parking lot equipped with one sensor node, which detects and monitors the occupation of the parking lot. The status of the parking field detected by sensor nodes is reported periodically to a database via the deployed wireless sensor network and its gateway. The database can be accessed by the upper-layer management system to perform various management functions, such as finding vacant parking lots, auto-toll, security management, and statistics report.

The information about the location and availability of a parking space near the destination is provided to the drivers by the current GPS-based vehicle navigation system, which is discussed by Pullola et al. [8]. They propose a scientific solution to parking allocation problem by utilizing the history and current status of the occupancy/availability. They model the availability of a parking lot and propose an intelligent algorithm that allows drivers to choose a parking lot with the maximum probability of getting it.

Rongxing et al. [9] propose a new smart parking technique based on vehicular communication for large parking lots. This technique permits minimizing the drivers' hassle and inconvenience. The proposed technique is characterized by employing parking lot RSUs to survey and manage the whole parking lot and is enabled by communication between vehicles and the RSUs. Once vehicles that are equipped with wireless communication devices, which are also known as onboard units, enter the parking lot, the RSUs communicate with them and provide the drivers with real-time parking navigation service and secure intelligent antitheft protection and friendly parking information dissemination.

In Banerjee [10], a new system based on image processing is proposed for the provision of parking information and guidance. The proposed system includes counting the number of parked vehicles and identifying the stalls available. The system uses images for detection of the vehicles. A camera for capturing the reference image is present at the entrance. For this purpose edge detection has been carried out using Prewitt edge detection operator and according to percentage of matching guidance and information, which is provided to the incoming driver.

Banerjee and Al-Qaheri [11] propose an intelligent and optimized scheme to solve parking space problem for a small city using a reactive search technique (named as Tabu search) assisted by rough set. Rough set is being used for the extraction of uncertain rules that exist in the databases of parking situations. The inclusion of rough set theory depicts the accuracy and roughness, which are used to characterize the uncertainty of the parking lot. Approximation accuracy is employed to depict accuracy of a rough classification according to different dynamic parking scenarios.

Geng and Cassandras [1] propose a smart parking system for an urban environment. The system assigns and reserves an optimal parking space for a driver based on the user's requirements that combine proximity to destination and parking cost while also ensuring that the overall parking capacity is efficiently utilized. This approach solves a mixed integer linear program (MILP) problem at each decision point in a time-driven sequence. The solution of each MILP is considered as an optimal allocation based on current state information and subject to random events such as new user requests or parking spaces becoming available.

The emerging of intelligent sensors results in the emergence and development of intelligent parking. Parking survey is one of the most important things for the parking managers and corresponding planners or researchers [12]. In this work, Chen et al. [12] discuss the problem of making a parking survey in intelligent parking systems where parking spaces, entrance, and exit are detected to acquire the occupation of the parking. They present three possible sensor layouts and corresponding algorithms to obtain the characteristic index needed in a parking survey. These intelligent parking systems can also do parking survey in different times and areas with the same detection data.

3. The proposed approach

In this paper, we will propose our approach to model a smart parking. This approach is based on a multi-agent system and genetic algorithms.

3.1 Multi-agent systems

An agent is a software system that is situated in some environment and that is capable of autonomous action in order to meet its design objectives [13]. In this work, we use the multi-agent system because it provides numerous advantages in the domain of allocation of vehicle into parking and drivers' comfort level. The proprieties of multi-agent systems offer autonomy to find the optimal parking, in real time. Thus, it provides a suitable framework for these systems. Also, they provide a number of important characteristics as the cooperation, negotiation, adaptation, and the mobility. That is, on the one hand, this autonomous agent perceives its environment, and on the other hand, the agent modifies its environment by its actions. Hence, an agent can dynamically find the adaptable parking in real time and according to the actual situation of the environment.

3.2 Genetic algorithms

Genetic algorithms are developed by Holland [14] to imitate the phenomena adaptation of living beings. They are an optimization technique based on the concepts of natural selection and genetics. It searches an optimal solution among a large number of candidate solutions within a reasonable time (the process of evolution takes place in parallel). Each of these solutions contains a set of parameters that completely describe the solution. This set of parameters can then be considered as the genome of the individual, with each parameter comprising of one or more chromosomes. They allow a population of solutions to converge step by step toward the optimal solution. To do this, they will use a selection mechanism of the population of individuals (potential solutions). The selected individuals will be crossed with each other (crossover), and some will be mutating by avoiding, whenever possible, local optima. The genetic algorithms are used primarily to treat both problems [15].

1. The search space is large, or the problem has a lot of parameters to be optimized simultaneously.

2. The problem cannot be easily described by a precise mathematical model.

We will combine multi-agent systems with genetic algorithms, for permitting the agent to choose the optimal actions. Therefore, our proposal is smart model which is based on the three following points:

1. Parking agent is a software agent which can manage, control the local optimization process, and exchange relevant information with neighboring agents.

2. Genetic patrimony which transformed between agents is used as inputs to the genetic algorithm. This genetic patrimony represents values of waiting and parking time and the distance between the vehicle and the proposed parking that are collected by sensor system.

3. Genetic algorithms are used to find the optimal solution for the current configuration; this is composed of two objective functions: one calculates the gain, and the other calculate the drivers' comfort level.

3.2.1 System architecture

The objective of our proposed system is to find the available parking according to the traffic road conditions. The proposed system determines the optimal parking for the vehicles circulated in the road network which need to use a parking. Thus, our system considers two mainly parameters: the number of vehicles and parking and environment data and waiting and parking time. The result of a system can be suggested to the driver or can be applied automatically as a part of a control system. In **Figure 1**, we present the architecture of our system. In this system, we have principally three agents: vehicle agent, parking agent, and agent profile. The parking agent can find the optimal parking placement according to

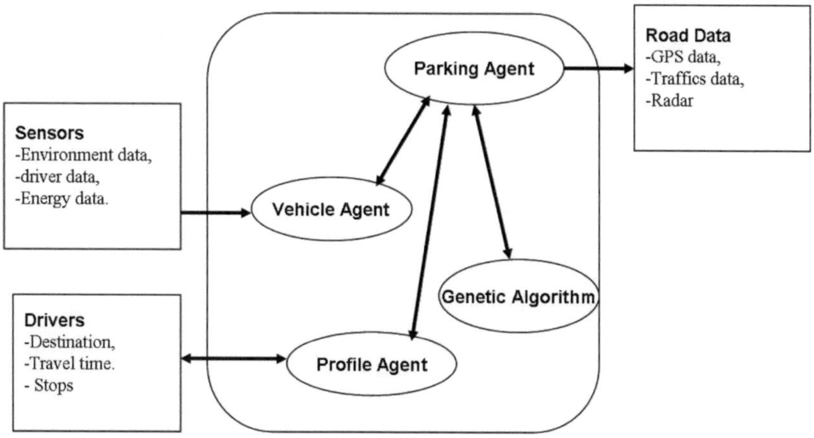

Figure 1.
Architecture of our system.

the profile of drivers' vehicle, such as distance between vehicle and proposed parking and waiting and parking time.

4. Parking agent

As we mentioned previously, each parking agent is able to manage, control, and regulate its parking environment to optimize the parking allocation and increase the efficiency and productivity of systems, thus increasing the profit. Therefore, the main objective of a parking agent is to solve the conflicts which can occur during the allocation of vehicles into the optimal parking. The drivers' comfort satisfaction is related to both the conditions of the environment and drivers' preferences over the system. In order to evaluate the drivers' comfort level, environmental parameters can be used as indices to form the function of drivers' comfort by using the actual value of the environmental parameters and drivers' preferences of these parameters. Therefore, the parking agent has been designed with three components for controlling and regulating its parking environment. These components are an optimizer, simulator, and comfort model.

4.1 An optimizer

It runs a genetic algorithm. Since heuristic algorithms have no guarantee to find the globally optimal solution within the limited iterations, in this research, GA runs 100 times in each time step to increase the possibility of achieving the global optimization, saving battery, and meeting passengers' preferences. In this principle, more runs of the optimization algorithm will lead to higher probability of achieving better results, but it will inevitably take more computation time. After many trials, it was found that 100 is a reasonable number of runs for balancing the solution quality and computation time cost.

4.2 Simulator

Each parking agent has a simulator that is used together to discover the drivers' comfort level and find the optimal parking in the prevailing conditions. The results of simulator could be optimized to achieve a satisfactory balance between discovery time and system performance. The optimizer repeatedly runs the vehicle flow simulations for every time and calculates the satisfaction of drivers' comfort level. The best drivers' comfort level is then used to generate the subsequent generation of general drivers' comfort level, and over a number of generations, the best candidates' comfort level is identified.

4.3 Comfort model

The drivers' comfort model permits to control the drivers' preferences via computer techniques to find the optimal parking which satisfies drivers' comfort and increases the efficiency and productivity of the system. In order to meet the compromise between system efficiency and drivers' comfort level, a parking agent needs to evaluate the number of vehicles in the parking and a drivers' comfort level that demand parking in response to maximizing the gain. However, drivers' comfort level and the cost of parking usually affect each other in an opposite way. Therefore, the main goal of a parking agent is to solve the conflicts between maximizing the gain and increasing drivers' comfort level. The drivers' comfort level is related to both the traffic road conditions and drivers' preferences over the environment. In order to evaluate the drivers' comfort in the vehicle, environmental parameters can be used as indices to form the function of drivers' comfort by using the actual value of the corresponding environmental parameters and drivers' preferences of these parameters. Generally, the distance between the vehicle and proposed parking and the waiting and parking time are used as parameters to evaluate the drivers' comfort level. Also, the cost and the gains of parking can be used as parameters to evaluate the revenues of the system.

5. Optimization process of parking allocation

As we say previously, the parking agent has an optimizer and simulator that are used together to discover the optimal parking that can satisfy the comfort level of drivers. The use of genetic algorithm has a major advantage over systems that rely on predefined values, as each parking agent enables a genetic algorithm to discover the optimal parking that may not resemble any predefined values, but may be optimal values for the current conditions of the road condition, as congestion. The optimizers should achieve a satisfactory balance between the revenue of solutions and drivers' comfort level. Thus, each parking agent executes a genetic algorithm to find the optimal parking that can be attributed to each system to perform optimal waiting time, increasing the drivers' comfort level.

5.1 Chromosomes' structure

To apply the genetic algorithm, we should define the genes. The gene can be characterized by its identifier and a set of values of parameters that can be applied to define the optimal parking that satisfies the drivers' comfort level. We use multiple forms in coding the genes. Firstly, we use the strings in encoding the identifiers of vehicle, and then we use real number for encoding the values of waiting and parking time and the distance between the vehicle and the proposed parking. **Figure 2** presents the structure of the gene.

To identify the best chromosome from the population, the optimizer runs a genetic algorithm with its different classic steps, as selection, crossover, and mutation. The parking agent has a simulator, which permits it to identify the best available solution from the population; the optimizer repeatedly runs the waiting time simulator for each vehicle in a given generation. After a number

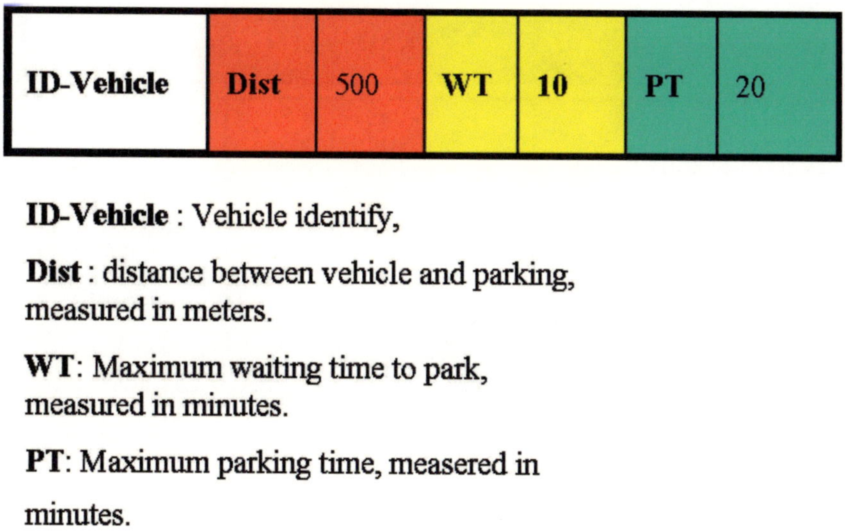

Figure 2.
Gene's structure.

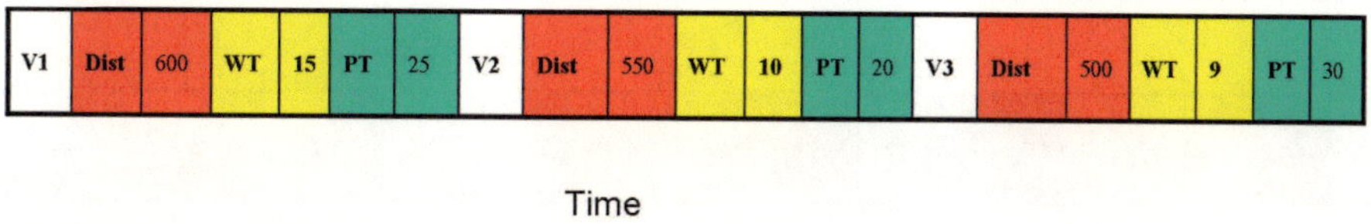

Figure 3.
An example of chromosome solution.

of generations, the best candidate values of parameters are identified. **Figure 3** shows an example chromosome of the parking agent.

5.2 Initialization, crossover, and mutation

Firstly, the initialization operator determines how each chromosome is initialized to participate in the population of genetic algorithm. Here, the chromosome is filled with the genetic material from which all new solutions will evolve [16]. This strategy means that the newly generated offspring may or may not remain within the new population, depending upon how they measure up against the existing members of the population [16].

Then, the crossover operator defines the procedure for generating a child from two parent chromosomes. The crossover operator produces new individuals as offspring, which share some features taken from each parent. The probability of crossover determines how often crossover will occur in each generation. In this approach, we will use the single-point crossover strategy which was adopted for all experiments. In this paper, the results for all experiments presented were generated using a crossover percentage of 50%, which is to say that at each generation, 50% of the new population were generated by splicing two parts of each chromosome's parents together to make another chromosome [16]. **Figure 4** presents the crossover operator.

Finally, the mutation operator will be applied. It defines the procedure for mutating the chromosome. Mutation, when applied to a child, randomly alters a gene with a small probability. In this work, the results presented here were generated using a 1% mutation probability, which was determined experimentally,

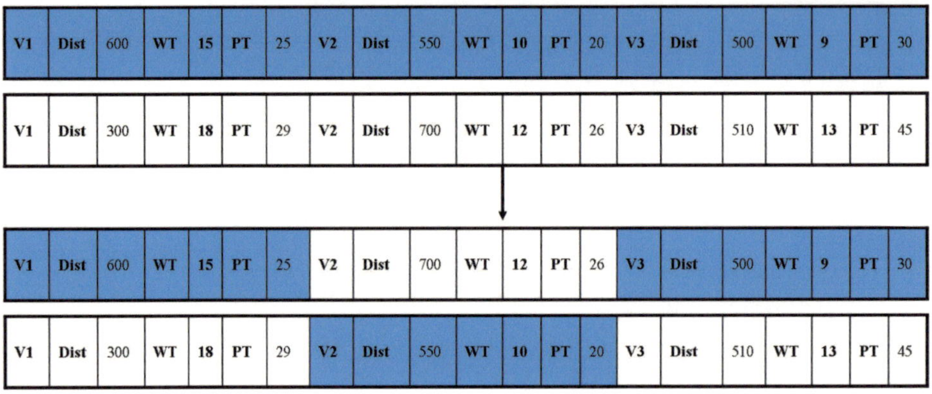

Figure 4.
The crossover operator.

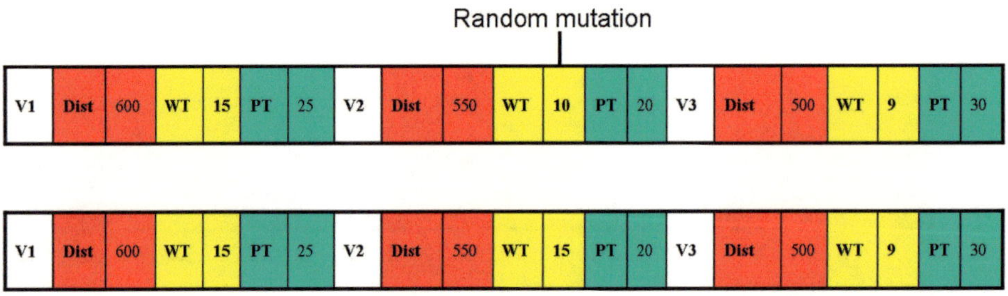

Figure 5.
The operator of mutation.

utilizing a single case of vector drivers' comfort level [16]. **Figure 5** presents the operator of mutation.

5.3 Evaluation of solutions

The purpose of evaluation system is to provide a measure for any given solution that represents its relative quality. In our resolution method for parking allocation problem for vehicles, the objective function used here works by calculating and summing the penalties associated with the waiting time, the parking time, and the distance between vehicle and proposed parking. Thus, we will use the objective functions to evaluate solutions of the parking allocation problem and examine the weighted relationship between the actual measured values of the waiting time, the parking time, and distance between vehicle and proposed parking. The objective functions used to evaluate solutions require a number of definitions that model the problem underlying structure, specifically.

- $V = \{V_1, V_2, V_3, \ldots\ldots V_n\}$ is the set of all vehicles in the road.
- $P = \{P_1, P_2, P_3, \ldots\ldots P_n\}$ is the set of all parking in city.
- *N1* and *N2* are the numbers of all parking in the city and all vehicles, respectively,
- WT_m, PT_m, and $Dist_m$ are the measured values of the waiting time, the parking time, and the closest distance, respectively.

- WT_c, PT_c, and $Dist_c$ are the comfort values of the waiting time, the parking time, and the closest distance, respectively.

- $[\mathrm{Dist}_{\mathrm{min}}, \mathrm{Dist}_{\mathrm{max}}]$ represents the distance comfort range. This range can be defined by drivers.

- $[\mathrm{T}_{\mathrm{min}}, \mathrm{T}_{\mathrm{max}}]$ represents the waiting time range.

Two important parameters are in our MAS-GA: the optimal assigned values of waiting and parking time and the closest assigned values to the distance between vehicle and proposed parking.

In this context, we have mainly two important functions $f(C)$ and $f(G)$ which permit evaluating the performance and efficiency of the proposed approach. These two functions are calculated by the parking agent.

The objective of this optimization mechanism is to minimize the value of $f(C)$ and maximize the gain $f(G)$ for evaluating the performance and the efficiency of our system. Firstly, we have:

$$f(C) = C_1 * {WT_c}/{WT_m} + C_2 * {PT_c}/{PT_m} + C_3 * {Dist_c}/{Dist_m} \tag{1}$$

C1, *C2*, and *C3* are the user-defined weighting factors, which indicate the importance of the three comfort factors and resolve the possible equipment conflicts. These factors take values in the range of $[0, 1]$. The drivers can set their own preferred values in different situations according to its travel. Since the travel period has a profound influence on energy savings, it should be taken into account in the control strategy design. Generally speaking, the parking agent activates the optimizer to tune the set point in order to obtain the acceptable drivers' comfort and to maximize the revenues [16]. Otherwise, the parking agent turns off all the resource lights and keeps the blind position to save energy if there are no vehicles in the parking. The objective function is defined in Eq. (1), and the optimization goal is to maximize these objective functions. Since the ratio between the measured value and comfort value is determined by drivers' play via graphic interface, it has an important role in achieving the control goal. Thus, it permits increasing the drivers' comfort level and optimize the parking places.

The second objective function permits controlling the gain of parking. The objective of this function consists of maximizing the total gain of parking. Thus, we can define this objective function as the following:

$$f(G) = N * \mathrm{Cost} \tag{2}$$

The cost represents the unit cost of parking.

6. Vehicle agent

When the vehicle needs to use parking, vehicle agents change their behavior and try to find a parking. This behavior of vehicle agent is triggered by a lower threshold E_1. The vehicle finds a parking; four parameters have to take into account: the actual position of the vehicle, the destination, the distance between the vehicle and the parking, and availability of place in the parking and the maximum waiting time. The proposed smart model should find the closest distance and minimize the waiting time.

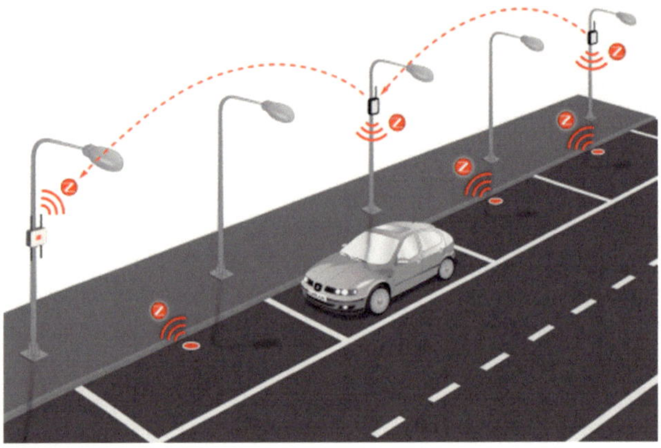

Figure 6.
Intelligent parking in the cities.

Firstly, each vehicle that needs to find a parking sends a request of parking to all the parking agents. Then, each parking treats this request and sends a response to the vehicle agent. Finally, the vehicle drivers are assumed to park its vehicle in the optimal parking. **Figure 6** presents an intelligent parking in the city.

7. Negotiation and cooperation

To model an optimal allocation of vehicle into parking, using the multi-agent system and genetic algorithm, we have to propose an efficient mechanism of negotiation, cooperation, and coordination between different agents. A single agent is unable to achieve some complex tasks, as the allocation parking problems because its capability is individually limited or although can complete, but its performance and efficiency are far lower than the performance and the efficiency with the cooperation and the coordination of many agents [17]. In order to solve charging station location conflict, the vehicle agents negotiate, each trying to find an available charging station. Therefore, when conflicts occur between two or many parking agents, it is important to limit their effects. In such case, negotiation techniques enable the involved parking agents to resolve their different conflicts by reaching compromises between the three parameters as the distance between the actual position of the vehicle, the waiting and parking time, and the availability of parking. This negotiation allows the parking agents to solve various conflicts at once and prevents new conflicts to appear. The parking agents negotiate with each other in order to decide whether the optimal plan can be applied to satisfy the need of drivers' vehicles in parking and in the same time reduce the waiting time of parking. Thus, in our proposed approach, the parking agents negotiate with each other in order to determine the best possible arrangement. This negotiation enables them to solve various conflicts simultaneously and avoid new conflicts from appearing. In **Figure 7**, we present an example of negotiation between two vehicle agents: vehicle agent 1 and vehicle agent 2. Thus, these two agents negotiate by proposing a plan of actions which can arrange the two agents. In each cycle, one parking agent makes a plan proposal to the other vehicle agents, which they can accept or reject. If they accept, negotiation ends; otherwise, the other parking agent makes a proposal at the next cycle.

Also, the cooperation is defined as the collaboration between vehicle agents or station agents to find the optimal action to solve conflicts in the research station

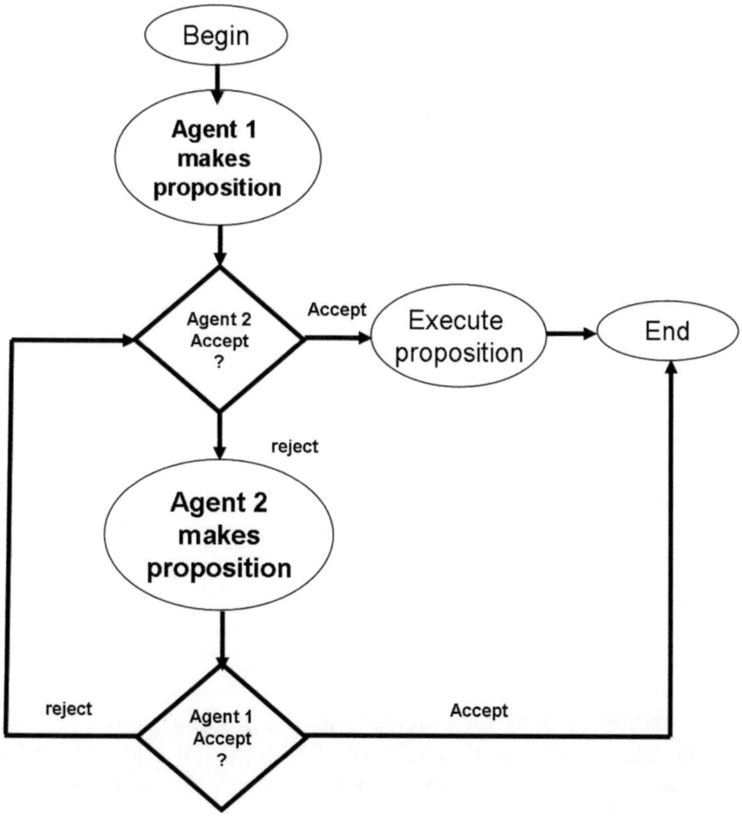

Figure 7.
Negotiation between two agents.

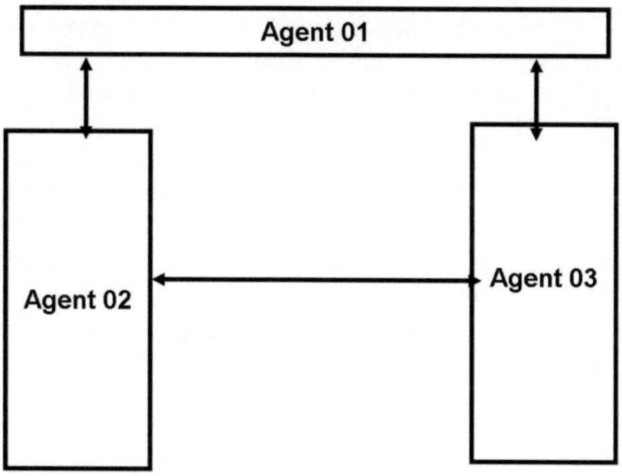

Figure 8.
Cooperation between agents.

problem. In this paper, we use cooperation in order to solve conflicts which occur between the different vehicles. Each parking agent cooperates with the other parking agents in order to find the optimal solution which permits reducing the waiting time and finds the closed parking. Each vehicle agent sends to the parking agent some data as the position, the remaining energy, and maximum waiting and parking time that permit to the parking agent analysis this data and performs its genetic algorithm. Next, each parking agent checks its list of requests to treat them and try to find the final optimal sequence of values of waiting time and the parking time that permit it to solve the conflict and avoid other conflicts to occur. The vehicle agents can accept or refuse the response of other agents according to their current situations. Thus, the parking agent sends the demand of the solution founded by its neighbors, and it waits to receive the responses of them, analyzes these responses, and determines whether the solution is possible or not. If a solution is feasible, it sends a confirmation to those of its neighbors that accepted this solution. **Figure 8** presents a simple configuration of cooperation between three agents.

8. Experiments and results

In this section, we present two case studies that illustrate how to design the different agents of our system and show collaboration between them. We use Jade to implement the different agents, vehicle agent, profile agent, and parking agent. Also, we use Java to implement the different steps of genetic algorithm as crossover operator, mutation operator, and the evaluation function.

The vehicle agent uses the sensor to learn the traffic data, waiting time, and parking time, which can be used as input in the genetic algorithms. The drivers can introduce their preferences in the profile agent via a graphic interface. The parking agent runs a genetic algorithm that can find the optimal values of waiting and parking time, which permit an optimal allocation of vehicle and increase drivers' comfort level.

During each simulation process, the optimizer of the parking agent can run a genetic algorithm. The optimizer computes the corresponding fitness value of each proposed parking according to the remaining energy of vehicle, the distance between vehicle and proposed parking, and waiting and parking time. In this experiment, the total number of parking place that can be served is set to 100.

When the vehicle agent sends a request to all station agents which are its neighbors, each parking agent can execute its genetic agent to define the optimal response, according to the parameters of vehicle agent as the distance, parking time demanded, and the maximum waiting time. This data is sent by the vehicle agent. It had the possible parking configurations distributed in the city. In this work, we use the Euclidean distance to measure the distance between the actual position of the vehicle and the demanded parking. Also, the parking agents can be cooperated and negotiated to define the final solution.

In **Figure 9**, we present the waiting time of vehicles when we use the approach without MAS-GA. We observe that the waiting time increases, according to the number of vehicles which demand parking.

When we use the approach of optimization with MAS-GA, each parking agent can interact with the other parking agents, and also it can perform its genetic algorithm to discover the optimal solution. We observe that the waiting time of vehicle decreases over the generation number. **Figure 10** shows the experimental results.

What is clear from **Figure 10** is that by using multi-agent system and genetic algorithm, the waiting time of vehicles is decreased in a few generations and in significantly less time than the approach without MAS-GA. For the MAS-GA approach, after 600 generations, the maximum expected waiting time achieved by Experiment 02 is less than 87 seconds, whereas the maximum expected waiting time achieved by Experiment 01 was rapidly augmented. Thus, the MAS-GA

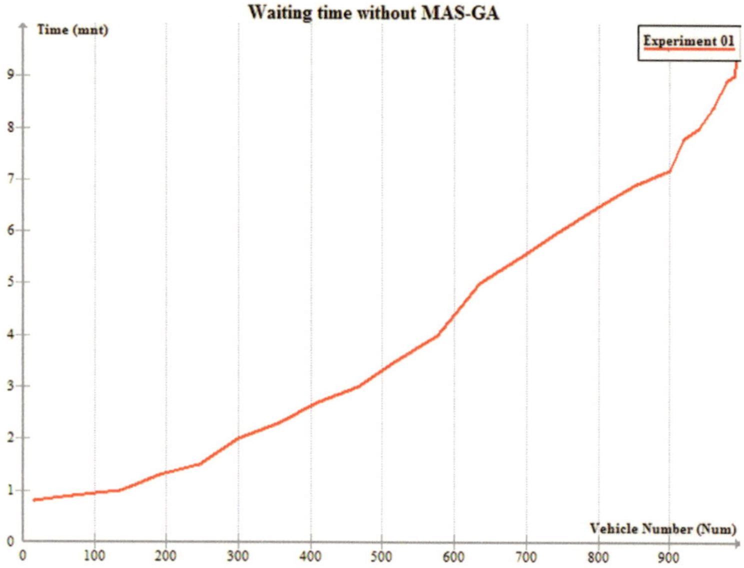

Figure 9.
The waiting time of vehicle without MAS-GA.

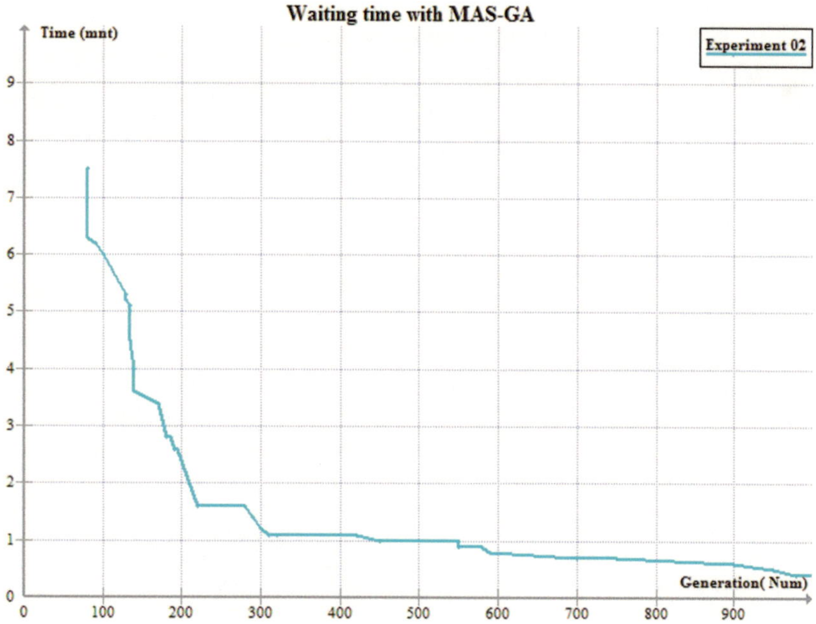

Figure 10.
The waiting time of vehicle with MAS-GA.

approach used in Experiment 02 converges to their optimal solution significantly faster than the other solutions presented in Experiment 01.

9. Conclusion

In this paper, we present a hybrid approach to control and manage an intelligent parking. Thus, we should find the available and adaptable parking according to the distance between the position of the vehicle and the proposed parking, the remaining energy of the vehicle, and waiting and the parking time. The experimental results show that our proposed approach can rapidly find the optimal parking compared with the other approaches. Also, when we use the MAS-GA approach, the revenues will be maximized.

Author details

Brahim Lejdel
University of El-Oued, El-Oued, Algeria

*Address all correspondence to: Brahim-lejdel@univ-eloued.dz
and lejdel.brahim@gmail.com

References

[1] Geng Y, Cassandras CG. A new smart parking system infrastructure and implementation. In: Proceedings of EWGT2012-15th Meeting of the EURO Working Group on Transportation. Vol. 54. September 2012. pp. 1278-1287

[2] Teodorovic D, Lucic P. Intelligent parking systems. European Journal of Operational Research. 2006;**175**:1666-1681

[3] Stéphane CKT, Alaoui EA, Cherif W, Hassan S. Improving parking availability prediction in smart cities with IoT and ensemble-based model. Journal of King Saud University-Computer and Information Sciences. 2020

[4] Diya T, Binsu CK. A genetic algorithm approach to autonomous smart vehicle parking system. Procedia Computer Science. 2018;**125**:68-76

[5] Li CC, Chou SY, Lin SW. An agent-based platform for drivers and car parks negotiation. In: 2004 IEEE International Conference on Networking, Sensing and Control. 2004. pp. 1038-1043

[6] Oyentaryo RJ, Pasquier M. Self-trained automated parking system. In: Control, Automation, Robotics and Vision Conference, 8th ICARCV; 6-9 December 2004. pp. 1005-1010

[7] Tang VWS, Zheng Y, Cao J. An intelligent car park management system based on wireless sensor networks. In: 2006 1st International Symposium on Pervasive Computing and Applications; 3-5 August 2006. pp. 65-70

[8] Pullola S, Atrey PK, El Saddik A. Toward an intelligent GPS-based vehicle navigation system for finding street parking lots. IEEE International Conference on Signal Processing and Communications; 24-27 November 2007. pp. 1251-1254

[9] Rongxing L, Xiaodong L, Haojin Z, Xuemin S. An intelligent secure and privacy-preserving parking scheme through vehicular communications. IEEE Transactions on Vehicular Technology. 2010;**59**(6):2772-2785

[10] Banerjee S, Choudekar P, Muju MK. Real time car parking system using image processing. In: 3rd International Conference on Electronics Computer Technology (ICECT). 2011. pp. 99-103

[11] Banerjee S, Al-Qaheri H. An intelligent hybrid scheme for optimizing parking space: A Tabu metaphor and rough set based approach. Egyptian Informatics Journal. 2011;**12**(1):9-17

[12] Chen NA, Wanga L, Jiaa L, Donga H, Lia H. Parking survey made efficient in intelligent parking systems, GITSS2015. Procedia Engineering. 2016;**137**(2016):487-495

[13] Wooldridge M, Jennings NR. Intelligent agents: Theory and practice. The Knowledge Engineering Review. 1995;**10**(2):115-152

[14] Holland J. Adaptation in Natural and Artificial Systems. Ann Harbor: University of Michigan Press; 1975

[15] DeJong K, Sarma J. In: Whitley D, editor. Generation Gaps Revisited, "Foundations of Genetic Algorithms 2". San Mateo: Morgan-Kaufmann Publishers; 1993

[16] Lejdel B, Kazar O. Using a Hybrid Approach to Optimize Consumption Energy of Building and Increase Occupants' Comfort Level in Smart City. Nature Switzerland AG: Springer; 2018

[17] Ferber J. Les systèmes multi-agents, vers une intelligence collective. Paris, France: InterEditions; 1995

Chapter 5

Alert Diagnostic System: SDA

Andressa dos Santos Nicolau, Marcelo Carvalho dos Santos, Victor Henrique Cabral Pinheiro, Filipe Santana Moreira do Desterro, Roberto Schirru, Mario Cesar Mello Massa de Campos and Alexandre Wanderley

Abstract

Currently, there is a trend in reduction of the number of industrial plant operators. The challenges are mainly during emergency situations: how to support operator time management without increasing operational risks? SDA focuses on this area and aims to increase operator situational awareness (ability to perceive, understand and predict the future behavior of a process) through new technological paradigms, such as Expert System and Ecological Human Machine Interface (HMI) in order to provide operational support, maintenance and optimization of refining, exploration and system of production of oil and gas plants. In SDA, the most critical alerts are shown by priority, along with decision trees, trend charts and variable comparison charts. SDA aims to assist control room operators in solving a critical problem in the oil industry, that is the loss of safety function, associated with alarms, during alarm flood. The SDA results of the SDA are presented through its implementation in Sulfur Recovery Units—URE, in the state of Rio de Janeiro, in Brazil.

Keywords: operator support, alarm floods, alarm processing, fault diagnosis, expert system, sulfur recovery units

1. Introduction

Industrial plants in general consist of a large number of integrated and interlinked process units. The information about the plant status is given by automated systems, which extract information of sensors spread over different parts of the process units and assist control room operators in making decisions and performing tasks to keep the plant operating in safe conditions and in an efficient way.

Control room operators are alert by automated systems through alarms. Nowadays, with digital technology, an alarm can be created within seconds and at almost zero cost. As a result, the number and frequency of alarms has increased significantly over the years. Alarms are typically set to a single operational state-triggered. Change of operating state, such as plant shutdown or plant startup may result in many alarms occurring at the same time. The amount of information presented is greater than as the human operator can actually perceive, so many of them are lost. This condition is called the alarm flood [1]. During alarm flood, operators can be overwhelmed by the large amount of alarms and not be able to keep the plant in safe operation condition posing a risk not only production process but also to the environment and human lives.

Alarm flood are one of the main causes of industrial plant accidents and cost millions of dollars each year, for example: Three Mile Island nuclear power plant accident in 1979, Esso Australia's gas plant explosion at Longford, in Vitoria, in 1998 [2], P-36 oil rig in the Campos Basin in Rio de Janeiro State, Brazil in 2001, that resulting in 11 deaths and total loss of the rig with an estimated financial loss of USD 400 million [3] and the Texas City Oil refinery explosion in 2005 [4] are clear examples of accidents problems that had contribution by alarm flood.

The system presented in this study is able to provide a real time support for control room operators in critical plant situations. Besides, it can assist the control room operators in time management and decision making process.

In order to present the methodologies developed and applied in SDA, this chapter will be divided into 3 items described below. Item 2 presents a brief description of the evolution of alarm systems in industrial plants. Item 3 presents SDA highlighting the methodologies implemented: expert system and ecological interface machine. And finally in item 4 is presented the conclusion of the chapter.

2. Alarm systems

Until the 1950s, an industrial plant control room was nothing more than a wall full of individual process indicators (panel), lights, switches, and moving pen charts. When something in the process was wrong one or more lights (alarms) came on in the panel along with beeps and indicate to operators which part of the process the problem was in. In this system introducing a new alarm was very expensive, where they had to be designed and implemented one by one in an often electromechanical system. With the evolution of computing many of these items became scarce and no longer met the needs of the operators.

In 1975 the first distributed digital control systems were created aiming to assist the industrial processes. These systems were responsible for record process variable data, calculate plant efficiency, assist in process monitoring and management and inform the operators about the plant state through alarms. In these systems the addition of a new alarm was done at no cost and within seconds. This led to a significant increase in the number of total alarms per operator, which can be verified according to **Figure 1**.

So increasing the amount of alarms configured did not bring more security to the process and not even facilitated operators' identification and decision-making. Because all events are treat as alarms and in situations of major plant disturbance are generated a huge amount of alarms, causing the so-called alarm flood. ANSI/ISA 18.2 [1] defines alarm flood as the occurrence of 10 or more annunciated alarms in any 10-minute period per operator. Since the alarms in these systems are presented

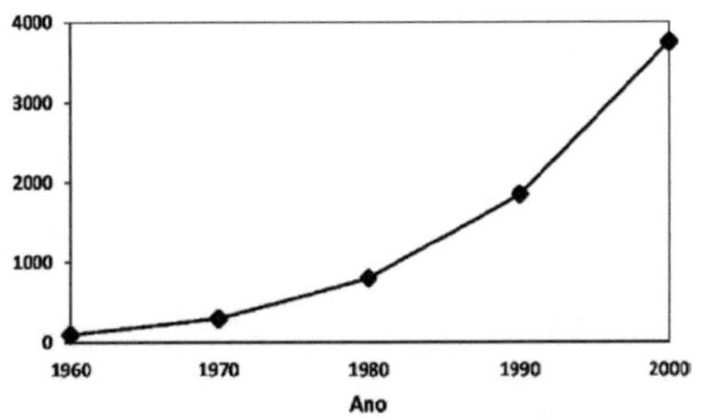

Figure 1.
Number of alarm per operator [5].

in sequence of events (SOE), it is impossible for the operator to understand them in a timely manner, causing often, due to a stressful situation, alarms very important for understanding the situation go unnoticed by the operators. In light of this problem, the major international engineering bodies came together to outline a set of methods, definitions and best practices for the design of an alarm system.

Studies on the most advanced alarm systems in operation [6–8] claim that a prioritization based on safety and plant urgency are the most frequently factors cited in advanced alarm systems literature and points out that the meaning of an alarm for control room operators depends on four factors: urgency, safety consequences, productivity consequences and relevance to the current task. It also cites the importance of the use categorizes by alarms through a time-based color coding available to operators and a dynamic severity rating. In other words, is important that the alarms are colored and segregated by urgency and ordered into categories by their severity.

The NRC alarm study [8] shows the effects on the performance of methods by which alarm processing results are disseminated to operational staff. The specific techniques analyzed in this study were suppression and dynamic prioritization. With suppression, minor alarms are not presented to operators but can be accessed upon request. In dynamic prioritization, the least important alarms are presented to operators, but differently from the most important ones. Because designers cannot anticipate every possible plant disturbance, some alarms may gain relevance in decision-making in a specific context. Thus, one of the advantages of dynamic prioritization cited in the study is that this approach does not omit any alarms to operators, unlike suppression.

According to the 191 standard British-based organization Engineering Equipment and Materials Users' Association (EEMUA) [9], alarm systems are an important means for automatic plant monitoring, drawing operator attention to significant process changes that require evaluation and action. They consist of field equipment, signal transmission, processing and visualization screen, being important tools to support the operator assisting in:

- Keep the plant process within a safe operating range. In this way, the operator is advised of potentially hazardous situations before the Emergency Shutdown System (ESS) is forced to intervene. This improves the plant assessment and helps to decrease demand from ESS, increasing the plant production and safety;

- Recognize and act to avoid situations that may lead danger to the plant. The role of ESS is intervening in a hazardous situation, however there may be cases where the plant deviates from its normal design operating conditions to a state in which the ESS is unable to act efficiently, such as during plant startup, which has a change of state.

- Identify deviations from operating conditions that could lead to financial losses;

- Understand the complex process conditions. Alarms can be an important diagnostic tool and are one of several sources of information that an operator can use during a critical process.

An alarm system is a crucial element in process plant operation, when well planned provide an additional layer of protection and can help operator prevent an abnormal situation from spreading, also offers benefits to the plant that include: increased safety, increased production, quality improvement and cost reduction.

3. Alert diagnosis system—SDA

SDA is a real-time operation support system that provides to control room operators an optimized flow of information from critical plant process variable changes, in order to assist decision-making in a short time, that can avoid operational situations like unexpected shutdown, loss of efficiency and even accidents. Thus, SDA is designed to support control room operators in the time management and take the right actions in order to keep the plant in a safe state.

The information provided on the SDA Human Machine Interface (HMI) is different from traditional alarm systems. In SDA alarms is called diagnostic alerts. The diagnostic alert can be formed by alarms already existing in the plant supervisory or not. They can be of the simple type (formed by a single alarm) or compound (formed by a set of alarms, which follow the order of logical operators, such as: and, or, between, less than, rate, etc.). Thus, the alert diagnostic not only identify changes in the process but identify problems (failures in the process), they by themselves already indicate the diagnosis of the situation. Besides, they are presented in a non-SOE approach, sorted by priority, where the highest priority alert will always occupy the top of the alarm list.

SDA is designed to monitor and process information about different process variables acquired and calculated by it, as well as process variables coming from different structures, such as: external databases, text files and excel spreadsheets. SDA consists of a data acquisition system, a knowledge based or rule bank (KB-SDA), a real-time diagnostic system, and a HMI. The real-time diagnostic system is an expert system, object oriented based on Artificial Intelligence Monitoring System (AIMS) technique [10]. **Figure 2** presented the SDA data flow.

For example, in SDA considering approximately 1000 (binary/analog) variables the time of one cycle is of 1 second. In other words, all data acquisitions (binary/analog) and the variables calculated from these are presented in a time less than 1 second. If a particular state or diagnostic alert is only dependent on the current values collected, it will be updated immediately, with a time interval of 1 second. If this state (or diagnostic alert) depends on a rate or time variation, it will be

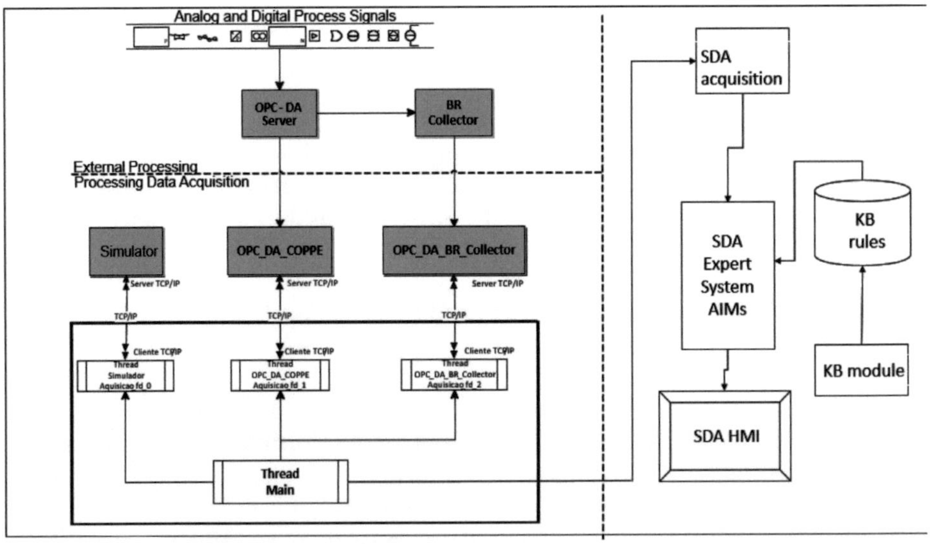

Figure 2.
SDA data flow.

accompanied by a 1 second step, until it reaches the required threshold. In other words, the operator response time is 1 second.

In this way, the SDA works at a level above the plant supervisory, where in alarm avalanche situations will show the highest priority alerts and the behavior of process variables associated with them giving the operator a clear plant situational awareness, quickly and offering benefits to the plant that include: increase security, increase in production, improve the quality, and cost savings.

3.1 Expert systems

Expert Systems or Knowledge Based Systems [11] were developed in the 1960s by the Stanford Heuristics Programming Project as a new intelligent method to find solutions for complex problems as a disease diagnosis. Edward Feigenbaum, widely known as the father of expert systems, defined it as "an intelligent computer program that uses knowledge and inference procedures to solve problems that are difficult enough to require significant human expertise for their solutions." In other words, an Expert System (ES) is a computational system that emulates the decision ability of a human expert in any topic.

A basic concept of ES is composed by a knowledge base – KB, where the intelligence of the system is stored, and an inference machine that process current facts based on the knowledge to generate new ones and conclusions (**Figure 3**).

The most relevant advantage in using ES is the independency between the KB and the inference machine. The KB can be changed or adapted to a new knowledge without the need of remodeling the inference engine. This capability makes this type of system a significant tool to handle diagnosis problems of many different types of power plants. ES are classified based on the paradigm in which information is represented in its knowledge base. The information can be represented on the Knowledge base in many forms: Logical Trees, Rules and Class [11].

The inference engine is responsible for join facts of a problem with the knowledge represented in the knowledge base, and establish new facts and conclusions. The information chaining process in the inference engine can be done in two ways: forward chaining and backward chaining. A forward chaining system begins with the facts initially known and uses the rules to draw new conclusions or take certain actions. All rules are checked to see if the initial facts satisfy some of them. Each satisfied rule is then fired, generating new facts that will be used to trigger other rules and so on until the problem is solved.

- Matching—where the antecedents satisfied by the facts are verified;
- Conflict Resolution—when more than one antecedent is satisfied, you must decide which of the rules will be fired first. This decision is called conflict resolution.
- Execution—in this step there is the execution of the rule, which can result in new facts as well as new rules.

In backward chaining, the inference process begins with choosing a solution and performs a search similar to depth searching [12, 13]. At first the known fact set is empty and as rules are fired this set becomes the set of facts that take the solution (object states). Thus, rules are triggered to generate values for object states or to generate intermediate facts that will later be used as a set of object state values.

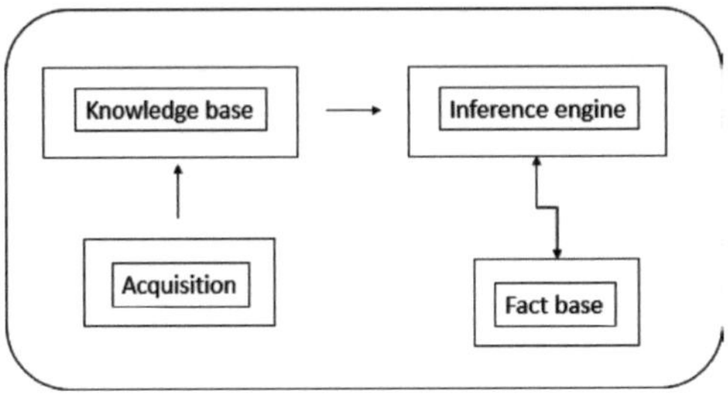

Figure 3.
A concept of expert systems.

3.1.1 SDA expert systems—AIMS

AIMS technology is a framework for developing real-time monitoring systems [14], which uses object-oriented (OO) concepts and expert systems. The AIMS kernel is the object-oriented knowledge-based (KB) expert system that acquires and calculates variables as well as their interdependencies, and maps them within a network of hierarchical objects, where rules are implicit in object operators and network topology.

The state of monitored variables updates a fact base, which is used by a real-time inference machine that activates and triggers knowledge base rules. A mail server is responsible for updating the operators on the HMI and manipulating the information.

The KB system has two main characteristics: (a) acquisition and maintenance of offline knowledge, which is built and modified through the KB Module and, (b) real-time monitoring representing the rules that define the real-time application in the which a network of hierarchical objects represents the rules that define the real time application, which is created according to KB rules and used by the inference engine.

The knowledge domain which comprises all monitored and calculated variables used by SDA, as well as their interdependencies, is mapped within a hierarchical structure of descended from parents' networks where each node contains a variable represented by an object which determines its attributes and operations. For each object is associated with a hierarchical level, which is used by the inference engine while firing the KB Rules. The lowest level is represented by the acquisition variables. **Figure 4** shows an example of a network of three level hierarchical objects.

In **Figure 4**, V22 means V22 is generated by V12. This hierarchical network represents all the rules contained in the KB rules that can be transformed into IF-Then structures, such as:

- IF (V11 updated or V12 updated)
- THEN (update V22 applying the operator considering its dependencies).

The KB module knowledge structure is based on five main class, shown in the hierarchical structure of **Figure 5**, where variables are abstract class from which the class are derived: analog (for representation of analog variables), binary (for representation of binary variables) and rate (for time variation representation).

The Message class represents the facts created from the inference of triggered rules. Every time the variable changes it generates a new fact (creating a new Message object). At the beginning of the monitoring process, the acquired variables are Message objects.

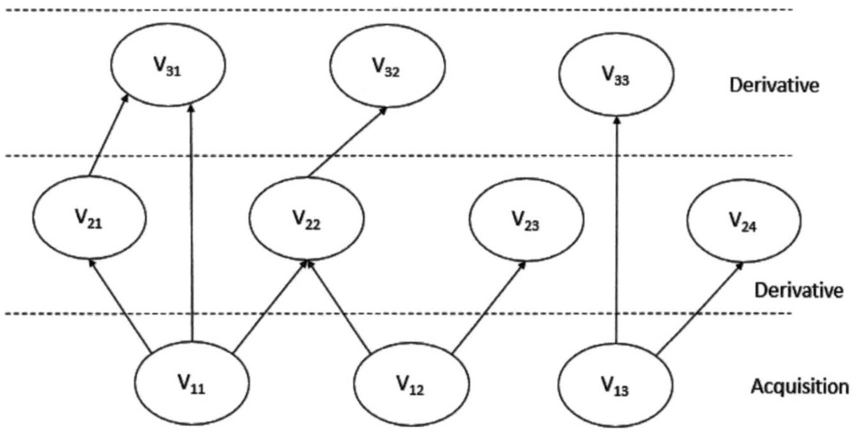

Figure 4.
Hierarchical network representation at AIMS.

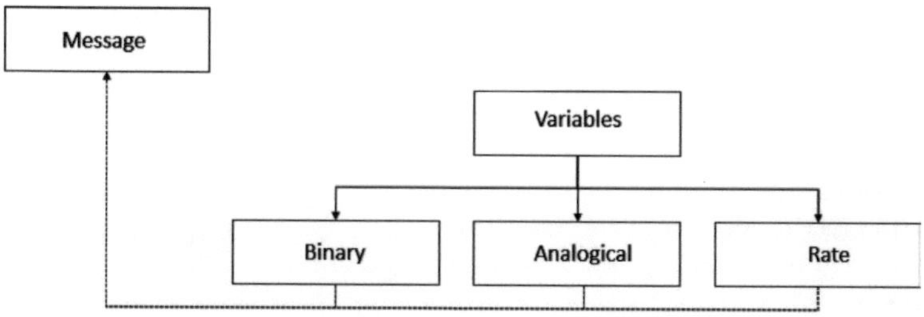

Figure 5.
Representation of hierarchical class in KB rules.

In the KB module the acquisition and maintenance of knowledge is done. In the KB module interface, the table shown in **Figure 6** allows the user to create object variables, edit their properties and operator rules as well as provide their dependencies on other variables (relatives and heritage).

Is not allowed to the users defines more than one link between two variables duplicate rules as well as inconsistent rules (different operations linked to the same antecedent with the same consequent). The rule problem without the condition side is eliminated due to the fact that the rule is made by linking two existing variables.

The track of time and the timing for rule activation is automatically done by the framework which does not allow descending nodes to have a refresh rate higher than their relatives' refresh rate. If more than one rule is activated by the inference machine, the conflict resolution strategy can be applied. The AIMS conflict resolution strategy takes as its first criterion the hierarchical level (in the object network) of the variables affected by the rule. Rules related to lower level variables have high priorities and will be trigged first. The inference process ends when there are no more rules to be activated.

Thus, in summary, the AIMS is able to receive data from binary and analog variables from an acquisition system, processing the data, performing calculations through logical and arithmetic operators and create new alarms (alarms not provided directly by the acquisition system).

The new alarms are called alert diagnostics and can be viewed as process alarms that require operator action, symptom alarms that indicate the failure of any unit component, or even prediction alarms that have the function of alerting the user that a failure could occur in the future if the current situation remains. Thus, the

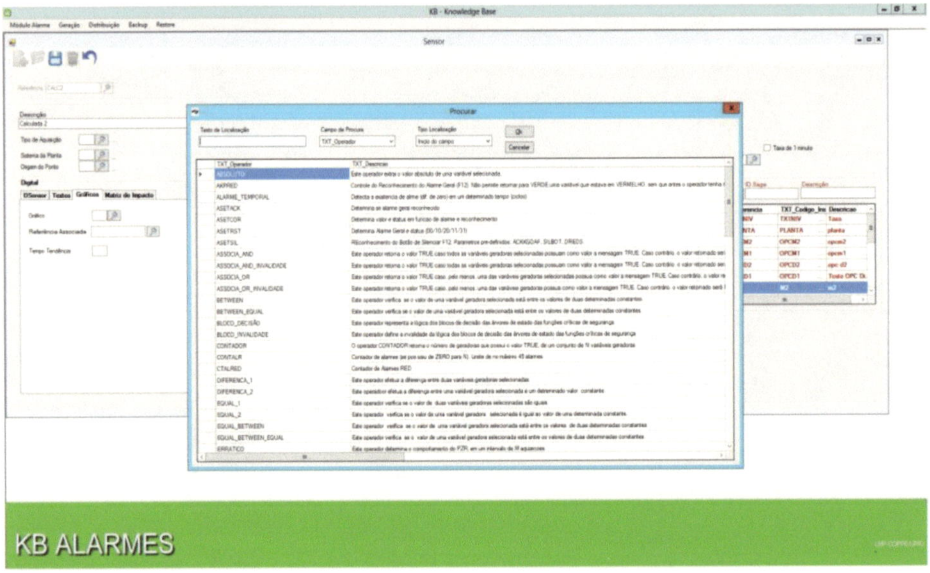

Figure 6.
SDA-KB module operators representation.

alert diagnostics generated by the SDA themselves represent a diagnosis of the current situation of the plant unit.

3.2 SDA HMI

SDA HMI concentrates the most important (critical) information in one place, preventing the "loss" of important alerts during emergency situations that cause alarm flood in traditional automation systems. Thus, it provides operators support in identifying the most critical problems, helping to prioritize actions and suppressing low criticality information. In normal operation, SDA also provides real time plant efficiency indicators, enabling actions to optimize (increase production) and increase reliability of operations, increasing operational continuity.

All information stored in KB SDA and processed in AIMS is presented to the users through the HMI. The frequency at which the information from HMI is refreshed is every 1 second. SDA HMI was developed based on the state of the art known as Ecological HMI [5, 6], where plant process variables (binary or analog) are represented by graphical objects which provide the behavior of these variables quickly and clearly, giving the operator a real time situational awareness of the plant. The graphic objects used in the SDA HMI will be: Plane Graph (x, y), Deviation Diagram, Sparkline, Radar Graphs, and Digital Diagram.

SDA HMI is divided into five different areas: (I) Date/hour, Binary Logical Annunciators Panel (ALB); (II) Ecological HMI; (III) Diagnostic alert list by priority; (V) Other options like: Ecological HMI of all diagnostic alerts, more diagnostic alert list, diagnostic sequence alert list, and additional information such as: instrument code, source, initial cause, action to be taken. **Figure** 7 shows the five areas in the SDA HMI with the ecological HMI option in area V. **Figure 8** shows the five areas in the SDA HMI with diagnostic alert list option in area V.

Diagnostic alerts are presented in HMI in three distinct areas:

a. in an area similar to the binary alarm annunciators (ALB), where diagnostic alerts are grouped by subsystems and by units (area I). **Figure 9** shows this information.

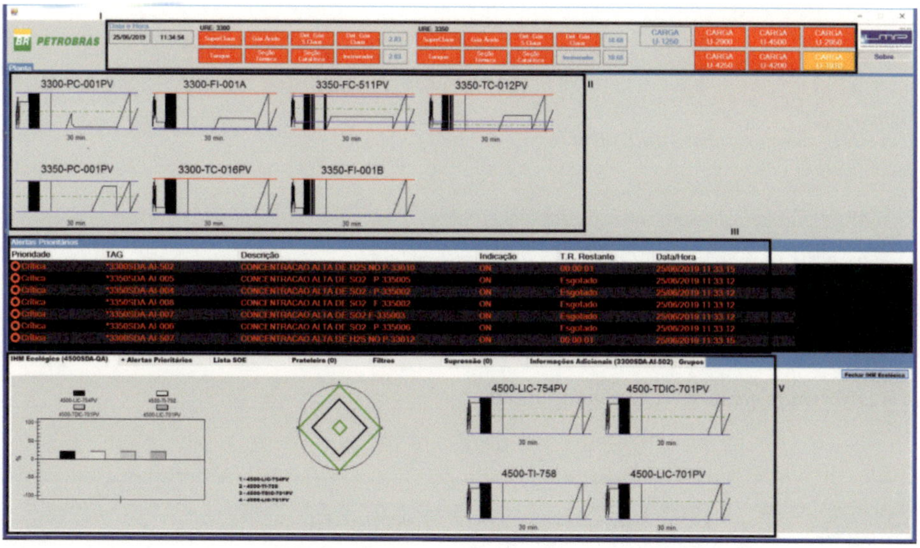

Figure 7.
SDA HMI with the ecological HMI option in area V.

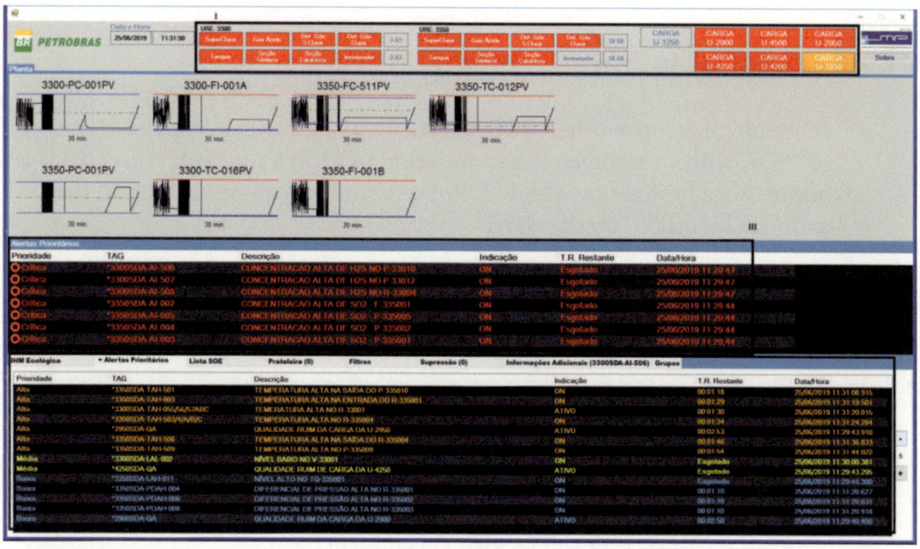

Figure 8.
SDA HMI with diagnostic alert list option in area V.

b. in a diagnostic alerts list sorted by priority: critical, medium, high, and low, displayed in red, orange, yellow, and blue, respectively (area III). In area III, only the seven more critical alert diagnoses are presented. (**Figure 10**). The diagnostic alerts list consists of:

- priority icon: icon that indicating the diagnostic alert priority;
- tag: diagnostic alert name;
- description: diagnostic alert description;
- on/off indication: indication whether the diagnostic alert is on or off;

Figure 9.
SDA HMI—diagnostic alert—ALB (area I).

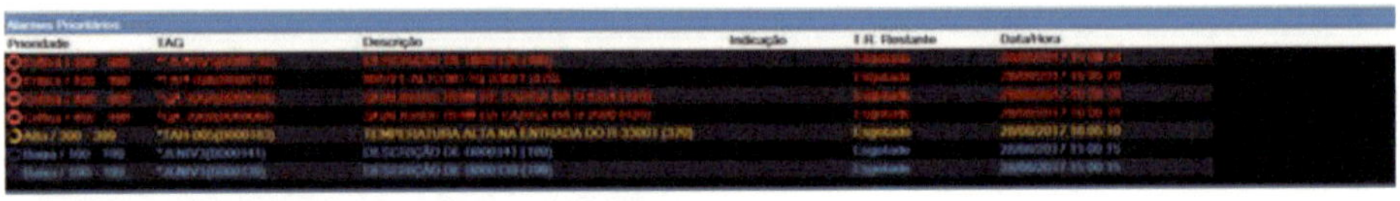

Figure 10.
SDA HMI—diagnostic alert list by priority (area III).

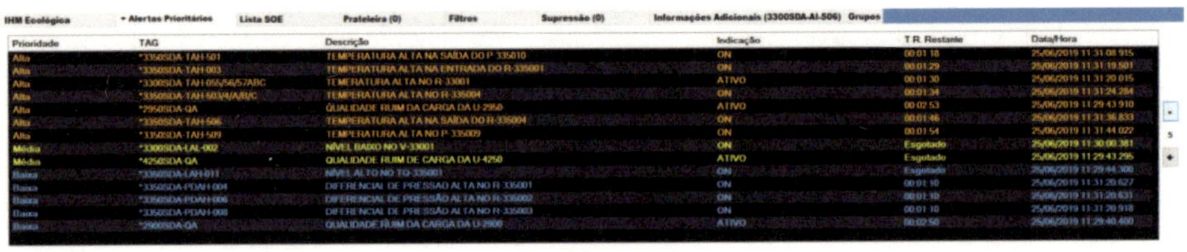

Figure 11.
SDA HMI—diagnostic alert list—more alerts (area V).

- remaining response time: In SDA each diagnostic alert is registered in the KB SDA with an associated response time. That is a time interval considered appropriate for the operator to take the necessary actions to correct the problem identified by the alert;

- activation data/time: date and time that the diagnostic alert was activated;

Important to point out that the most important diagnostic alert will always occupy the top of the list due the dynamic prioritization method.

c. and in another priority list in area V. The list aims to show more diagnostic alerts by priority, 8th onwards, not shown in area III. The list consists of: priority icon, tag, description, on/off indication, remaining response time and activation data/time. **Figure 11** shows this information.

3.3 SDA implementation

SDA was designed to be implemented in any onshore and offshore production system. Currently, SDA is implemented in two Sulfur Recovery Units in the state of Rio de Janeiro, Brazil with the following results:

1. Presents the quality and importance/priority of the seven strands that compose the load of the Units, all as technical documentation available online and in real time at SDA. In case that the operator does not realize in short time that the streams quality is not good, and does not quickly deviate this streams, it will generate deposition and lead to the unit shutdown, causing financial losses and increased emissions for the environment. This information is always available from the ecological HMI, and the operator is able to check the deviation quickly by looking at the SDA HMI;

2. Presents all diagnostic alerts regarding the prioritization of H_2S, SO_2 and NH_3 detectors from the all Units area. These warnings are critical and prevent the operator from going into an area with the presence of gas that could endanger their health and physical integrity;

3. Presents the diagnostic alert prioritization of the critical unit process stored into the KB SDA which has been built on the operational experience of operators and design engineers and HAZOP information's and others. This knowledge was previously elaborated, considering various scenarios, in order to assist the operators in decision-making in abnormal situations;

4. Presents, in real-time, information about sulfur removal efficiencies from Units and their subsystems. This information enables operators to act to minimize the tendency of units to lose efficiency;

5. Monitors critical controls on manual, active alarms and bypass, etc.;

6. Perform process and unit equipment diagnostics: vessels, distillation towers, heat exchangers, ovens, compressors, etc.;

7. Follows the desired limits or operation regions (pressure, temperature and composition) of the processes and equipment, ensuring that they are respected not only during normal operation but also during plant starts and plant stops;

8. It considers the different unit operating modes (starts, stops, steady state, disturbances, etc.) to generate diagnostics alerts specific to each situation;

Early tests with SDA show that concentrating all critical information's about the process on a single HMI helps the operator in time management. Preventing him from consulting different tools and screens to be situational awareness and take decision. In addition, the flexibility to create diagnostic alerts enables monitoring of situations not provided by the traditional alarm management system.

Tests applied in order to evaluate the operator situational awareness (ability to perceive, understand and predict the future behavior of a process) showed that in normal operation using traditional automation systems the operator situational awareness was around 50% and dropped during emergency situations for values below 20%. With SDA, during emergency, the operator situational awareness was around 55%, that was higher than those obtained in normal operation in traditional automation systems. This demonstrates the importance of an Intelligent Real-Time Decision Support System for oil industry control room operators aim to anticipate failures and specially to avoid risk situations [15].

SDA is also being evaluated for use on offshore production platforms to increase the safety and reliability of these processes. This system can be implemented in any industrial process to providing an intelligent system, coupled with an ergonomic HMI that allows operators to reduce your cognitive load in search of the root cause of a certain problem.

4. Conclusion

SDA was developed to support operators in time management in case of emergency situations providing optimized plant process information in order to increase operator situational awareness, so that they make appropriate decisions and actions aiming to increase the safety, integrity and reliability of plant processes and equipment.

SDA HMI aims to increase operator situational awareness by explaining why diagnostic alerts through decision trees, trend graphs, etc. and also by presenting recommendations for mitigating actions. The aim is to help operators identify process fails quickly and prevent the abnormal situation from spreading, increasing risks to people, equipment and the environment.

The methodologies developed and implemented makes SDA a potential support system from easy applicability in different onshore and offshore from oil industry, mainly impacting the sector in reduction of the number of disturbances and shutdowns, production and increased plant safety. Besides that, SDA impacts in relation the solution of alarm flood, anticipate failures, avoid hazardous situations, avoid misdiagnosis, prevent breakdown and /or equipment unavailability, prevent leaks, avoid emissions (e.g., torch) and debris. This increases the safety, reliability and production of complex oil industry processes. A good indication of this impact was the installation of SDA in Sulfur Recovery Units in Brazil.

Concluding we can say that SDA is a new safety barrier providing diagnostics alerts that can prevent an abnormal situation from evolve into accidents and may decrease the chance of human mistake, especially in identifying the root cause, which if misdiagnosed, may lead them to take wrong actions that increase the consequences of incidents, accidents or plant shutdowns.

Acknowledgements

The authors would like to acknowledge: CAPES (Coordination for the Improvement of Higher Education Personnel), CNPq (National Council for Scientific and Technological Development) and Petrobras for financial support.

Author details

Andressa dos Santos Nicolau[1*], Marcelo Carvalho dos Santos[1], Victor Henrique Cabral Pinheiro[1], Filipe Santana Moreira do Desterro[1], Roberto Schirru[1], Mario Cesar Mello Massa de Campos[2] and Alexandre Wanderley[2]

1 Federal University of Rio de Janeiro, COPPE/UFRJ, Rio de Janeiro, Brazil

2 CENPES/ PETROBRAS, Rio de Janeiro, Brazil

*Address all correspondence to: andressa@lmp.ufrj.br

References

[1] ANSI/ISA 18.2-2016: Management of Alarm Systems for the Process Industries. 2016. Available from: https://www.isa.org/store/ansi/isa-182-2016/management-of-alarm-systems-for-the-process-industries/46962105 [Accessed: 30 August 2019]

[2] Hopkins A. The Esso Longford Gas Plant Accident—Report of the Longford Royal Commission. 1999. Available from: https://www.parliament.vic.gov.au/papers/govpub/VPARL1998-99No61.pdf Nª 61, Session 1998-99, Government Printer for the State of Victoria [Accessed: 30 August 2019]

[3] Figueiredo MG, Alvarez D, Adams RN. Revisiting the P-36 Oil Rig Accident 15 Years Later: From Management of Incidental and Accidental Situations to Organizational Factors. 2018. Available from: http://www.scielo.br/pdf/csp/v34n4/en_1678-4464-csp-34-04-e00034617.pdf [Accessed: 30 August 2019]

[4] Isiadinso CC. BP texas city refinery disaster accident and prevention report. ResearchGate. 2015;**1**:1-9. DOI: 10.13140/RG.2.1.2317.4569

[5] NAMUR, Current NAMUR Recommendations (NE) and Worksheets (NA): Alarm Management. 2008. Available from: http://www.namur.net/en/recommendations-and-worksheets/currentnenahtml [Accessed: 30 August 2019]

[6] Hollifield B, Habibi E. Handbook of Alarm Management: A Comprehensive Guide. 2nd ed. Isa. 2006. ISBN 10: 193600755X. ISBN 13: 9781936007554

[7] O'Hara JM, Brown WS, Lewis PM, Persensky JJ. Human-System Interface Design Review Guidelines. NUREG-0700 Rev. 2. 2002. Available from: https://www.nrc.gov/docs/ML0217/ML021700337.pdf [Accessed: 30 August 2019]

[8] Brown WS, O'Hara JM, Higgins JC. Advanced Alarm Systems: Revision of Guidance and Its Technical Basis, NUREG/CR-6684. 2000. Available from: https://www.nrc.gov/docs/ML0037/ML003770903.pdf [Accessed: 30 August 2019]

[9] EEMUA, 191 Alarm Systems—A Guide to Design, Management and Procurement. 3rd ed. 2013. ISBN: 0-85931-192-2

[10] Schirru R, Pereira CMNA. A Real Time Artificially Intelligent Monitoring System for Nuclear Power Plants Operators Support. Netherlands: Kluwer Academic Publishers; 2004. Available from: https://link.springer.com/content/pdf/10.1023%2FB%3ATIME.0000019127.50572.9b.pdf [Accessed: 30 August 2019]

[11] Novák V, Perfilieva I, Mockor J. Mathematical Principles of Fuzzy Logic. Springer US, New york: Kluwer Academic; 1999. 320p. Springer. ISBN: 978-1-4615-5217-8

[12] Augusto JPSC. Sistema Inteligente de Suporte ao Diagnóstico de Angra I [thesis]. Universidade Federal do Rio de Janeiro, Engenharia Nuclear, Escola Politécnica; 2017

[13] Gonzales AJ, Douglas DD. The Engineering of Knowledge-based System—Theory and Pratice. Hemel Hempstead: Prentice Hall International; 1993. 523p. DOI: 10.1017/S0263574700017252

[14] Machado L. Modelagem do Conhecimento para sistemas inteligentes de Monitoração de Segurança em Tempo Real para Usinas Nucleares [thesis]. Universidade Federal do Rio de Janeiro, Departamento de Engenharia Nuclear; 2014

[15] Satuf E et al. Situation awareness measurement of an ecological interface designed to operator support during alarm floods. International Journal of Industrial Ergonomics. Elsevier. 2016:179-192. DOI: 10.1016/j.ergon.2016.01.002

Section 2

Application in Electrical Power Systems

Chapter 6

Intelligent System for the Estimation of Gases Dissolved in Insulating Mineral Oil from Physicochemical Tests

Ivan Nunes da Silva, Rogério Andrade Flauzino, Danilo Hernane Spatti, Renato Pagotto Bossolan and Bruno Augusto Trevisam

Abstract

The objective of this work was to make the modeling through artificial neural networks of the gas concentrations dissolved in insulating mineral oil from the results of physicochemical tests. In this case, a mapping between the data of physicochemical tests and gas chromatography was obtained by means of artificial neural networks. The proposed approach proved to be efficient to identify the amount of gases, taking the following attributes as input: color degree, density, dielectric rigidity, interfacial tension, power factor of the insulating oil, neutralization index, and water level. In addition, artificial neural networks provide not only a new methodology to support decisions but also satisfactory results comparatively to actual analyses when referring to the estimation of gases.

Keywords: intelligent systems, artificial neural networks, power transformer

1. Introduction

The power transformer is an extremely important and expensive equipment consisting, in short, of two or three windings, the core, and the insulation system. The windings are arranged such that the magnetic leakage flux is dispersed as little as possible. The core is the medium by which the magnetic flux finds a path of low reluctance and preferably flows through it. Among these elements, core, windings, and the transformer tank, there is also the insulation system. The insulation system consists essentially of insulating paper and insulating mineral oil.

During its operation, various wear and aging processes may occur in the transformer insulation system. Examples can be hot spots, overheats, and overvoltages, among other phenomena that change the transformer insulation system [1]. Thus, the reduction in the lifetime of transformers is directly linked to the deterioration of the dielectric materials used in their manufacture. In this sense, there is an important motivating aspect for the development of supervision and preventive maintenance programs in order to provide an increase in the useful life of the equipment and, consequently, a better management of them under the command of the electric

IntechOpen

utilities. In fact, the electricity market is becoming increasingly competitive, and the high costs involved in maintenance require the development of processes in order to lower its costs, increase asset performance, as well as extend its useful life [2].

Among the methodologies used to identify failures in power transformers immersed in insulating mineral oil, we highlight those based on the monitoring of their electrical parameters, acoustic monitoring, and the evaluation of dissolved gas concentrations in oil. On the other hand, the condition associated with the insulating paper present in the power transmission transformer coils is one of the main responsible for the lifetime of this equipment. Winding insulation paper can deteriorate abruptly, resulting in unpredictable failure of these transformers, particularly if in contact with oxygen, humidity, and metal contaminants. All these possible scenarios can be avoided if the degradation process is discovered in time [3].

For identifying failures from the dissolved gases in the oil, several chromatographic tests are then performed, which result in their respective concentrations. The examination of such dissolved gases is based on the premise that failures from partial discharges, high energy discharges, corona effect, overheating, etc. react with the insulating mineral oil causing alterations in their physicochemical properties, which are responsible for the release of gases that eventually react with the oil and, consequently, are dissolved within it. Therefore, the dissolved gas analysis (DGA) in insulating mineral oil is one of the main methods used to diagnose the insulation situation of the power transmission transformer windings. In addition, proper interpretation of DGA results is one of the most significant procedures for detecting fault types, as well as for identifying the process of degradation of its insulation [4, 5].

In addition to the chromatographic tests, with which it is possible to obtain the volumetric values of the main dissolved gases in the oil, there are also physicochemical tests that allow to identify several physical and chemical characteristics present in the insulating mineral oil. An important feature associated with physicochemical tests is that they can all be performed at the location where the transformers are installed, and part of them can be performed with the equipment in service, i.e., without interruption in the supply of electricity. Moreover, when power transmission transformers are in service, they may be subject to a number of environmental factors, such as heat and humidity, which must lead to an increasing aging process over time. Consequently, the electrical properties of the insulating mineral oil (resistivity, dielectric losses, relative permittivity, and breakdown strength), as well as the chemical (acidity and humidity) and mechanical properties (tensile strength and viscosity), are also modified, which directly impact its useful life [6, 7].

Therefore, it is emphasized here the motivation of a system for estimation of gases dissolved in oil from the results of physicochemical tests. Thus, the objective would then be to adapt existing methodologies based on the concentration of dissolved gases to, based on physicochemical tests, predict possible incipient failures in power transformers.

Since the estimation of these results is a complex problem, which does not have a mathematical formula capable of relating both tests, an interesting alternative is the use of computational intelligence techniques. Among these techniques, we can highlight the artificial neural networks, which do not require detailed knowledge about the relationship between the input space and the output space, as they are expert in dealing with nonlinear mappings [8].

It should be emphasized that the great novelty of this work is the precise estimation of the concentration of gases that are dissolved in insulating mineral oil, through neural networks, from the results of physicochemical tests. For the estimation of gases, we then use data from physicochemical tests performed on insulating oil taken from transformers in operation. The developed technique allowed the estimation of these gases without the need to perform the chromatographic test. Consequently, the amount of power outages will be considerably reduced, and the associated costs in determining the gas concentration in the mineral oil will follow the same downward trend.

For this purpose, the organization of this article was divided in five sections. In Section 2, several aspects related to the main experimental tests performed on the insulating mineral oil are described. In Section 3, the intelligent system based on artificial neural networks for mapping the relationship between both physicochemical and chromatographic tests is presented. The main results produced by intelligent system in order to validate developed methodology are reported in Section 4. In Section 5, we present the conclusions about the applicability of the proposed technique.

2. Aspects related to the experimental tests in insulating mineral oil

Depending on the type of petroleum used for fractional distillation, the insulating mineral oil may have paraffinic base (paraffinic oil) or naphthenic base (naphthenic oil). Its classification is based on the results of the percentage of paraffinic (PC), naphthenic (NC), and aromatic (AC) carbons. This classification can be obtained by using infrared spectroscopy techniques by determining the amount of paraffin carbons. Oils with PC lower than 50% are considered naphthenic, while those with PC equal to or greater than 56% are classified as paraffinic; between 50 and 56% are intermediate oils.

When different mineral oils are mixed, the properties of the resulting mixture will be an average of the properties of each of the components, since the mixed oils are of good quality. If one of these components is of poor quality, the resulting oil will be of poor quality.

Determining the physicochemical properties of insulating mineral oil is of fundamental importance to ensure the operating conditions of transformers and to maintain or extend the lifetime of these equipments [9, 10]. In order to prevent premature degradation, which can be translated into financial loss, it is therefore necessary to monitor the physicochemical properties of the insulating mineral oil. Through the analysis of physicochemical characteristics, it is then possible to evaluate the quality of both the oil and the transformer itself. These analyses are of great importance for the maintenance of electrical equipments, as they allow planned interventions to correct any defects and, in some cases, early failure diagnoses [11, 12].

To test the insulating mineral oil, it is necessary that the transformer is working, because in operation the oil goes through an aging process. This aging is due to the demand for temperature increase, to the action of oxygen, and to contact with materials present in its construction, resulting in deterioration of oil properties. This deterioration generates by-products that promote its acceleration, that is, it generates a chain reaction where the insulating mineral oil loses its insulating properties and, as a result, the cellulose degrades generating sludge. The process that governs the oxidation of hydrocarbons is the peroxidation mechanism (peroxide formation), where there is the formation of intermediate products, which may be alcohols, aldehydes, and ketones depending on the species that originated them.

In **Table 1**, it showed the main experimental tests and their respective standards, which are usually recommended to trace the transformer isolation conditions. Therefore, the various tests performed on the insulating mineral oil in use allow to diagnose some problems, such as hot spots, overheating, and leaks, as well as informing about the insulating and thermal quality of the respective oil.

Tests	Technical standards
Aniline point	ABNT–NBR–11343 [13]
Chromatography analysis	ABNT–NBR–7070 [14, 15]
Cinematic viscosity	ABNT–NBR–10441 [16]
Color	ASTM–D1500 [17]
Density	ABNT–NBR–7148 [18]
Dielectric rigidity	IEC–156 [19] and ABNT–NBR–6869 [20]
Factor of losses at 90°C	IEC–247 [21]
Index of total acidity	ASTM–D974 [22]
Interfacial tension	ABNT–NBR–6234 [23]
Oxidation stability	ABNT–NBR–10504 [24]
Point of splendor and combustion	ABNT–NBR–11341 [25]
Polymerization level related to insulating paper	ABNT–NBR–8148 [26]
Tenor of water	ABNT–NBR–5755 [27]

Table 1.
Experimental tests and technical standards applied in insulating mineral oil.

On the other hand, chromatographic tests, also known as dissolved gas analysis (DGA), are diagnostic tools for the detection and evaluation of incipient failures in insulating mineral oil. When a transformer failure occurs, the insulation system will provide chemical degradation, which will generate the production of various gases that will be present in the oil. These concentrations of gases are related to a type of transformer failure.

The main gases analyzed by DGA are listed in **Table 2**. In normal operation, the presence of these gases in the transformer is also normal, i.e., it does not indicate the existence of failures. The only exception is carbon dioxide (CO_2), whose presence indicates a failure inside the equipment.

A transformer failure can result in the variation of the standard concentration of these gases in the oil. The increase in the gas concentration in the transformer indicates the occurrence of a failure that results in the insulation oil saturation. With this saturation, gas is released from the oil by modifying the characteristics of

Dissolved gas	Chemical formula
Acetylene	C_2H_2
Carbon dioxide	CO_2
Carbon monoxide	CO
Ethane	C_2H_6
Ethylene	C_2H_4
Hydrogen	H_2
Methane	CH_4
Nitrogen	N_2
Oxygen	O_2

Table 2.
Gases evaluated by DGA technique.

the transformer. The amount of gas released depends on the oil temperature and the type of gas. Such gas production can be classified into three groups, namely, polarization, corona effect, and electric arc. This classification is based on the severity with which power is released during the failure. The largest and smallest amount of energy released is associated with the electric arc and corona, respectively [28].

In relation to the polarization, the gases released in the oil at low temperature are methane and ethylene, while at high temperature are methane, ethylene, ethane, and hydrogen. In cellulose, the gases generated at both high and low temperatures are carbon monoxide and dioxide. Regarding the corona effect, the gas produced in the oil is hydrogen, and the gases released by cellulose are hydrogen, monoxide, and carbon dioxide. For the electric arc, the gases released in this case are hydrogen, methane, ethane, ethylene, and acetylene.

Related to temperature, laboratory research shows that CO, CO_2, and water originate when cellulose is overheated to a temperature of 140°C. Pyrolysis, which is destruction at temperatures above 250°C, produces more CO gas than CO_2. In this case, there is also the formation of water, coal, and tar. At temperatures above 500°C, methane (CH_4), ethane (C_2H_6), ethylene (C_2H_4), CO_2, and water originate when O_2 is present.

For the extraction of gases from the insulating mineral oil to be correct, it is then necessary that there are no bubbles in the collection vessel. The sample should be slightly warmed or injected into a degassing apparatus. The sample is subjected to vacuum, and the gases are collected in a graduated burette, which is taken to the chromatograph. The chromatograph is basically an apparatus used for chemical analysis of substances, capable of separating the various components of the sample.

Thus, from values obtained in physicochemical tests performed on the insulating mineral oil, we have then developed an intelligent system capable of estimating the gas values present in the insulating oil of the transformer.

3. Artificial neural network for estimating gases dissolved in insulating mineral oil

Computational models based on artificial neural networks are inspired by the knowledge we have about the nervous system of humans, which has the ability to learn from experience. They are composed of artificial neurons, which are also interconnected by artificial synapses.

Artificial neural networks can be applied to various engineering- and science-related problems. One of the areas with the greatest potential for their applicability is the universal approximation of functions, whose purpose is to represent functional relationships between inputs and outputs of typically nonlinear systems.

They can be applied to solve problems coming from various areas of knowledge, such as engineering, medicine, chemistry, and physics, where the possible solutions to the problems presented are difficult to obtain by conventional techniques.

One of the most commonly used architectures in artificial neural networks is that known as multilayer perceptron (MLP) [8], which will be also used in this application to map the relationship between results from physicochemical tests and those produced by chromatographic tests.

In addition, MLP networks are also characterized by the high possibilities of application in various types of problems related to the most different areas of knowledge, which is also considered as one of the most widely used in terms of applicability.

Figure 1 illustrates an MLP topology consisting of three neural layers, which has n signals in the input layer, with n_1 neurons (first layer), n_2 neurons (second layer), and m neurons in its output layer..

Each of the neurons of this topology illustrated in **Figure 1** can be represented according to the terminology adopted in **Figure 2**.

In purely mathematical terms, the internal processing carried out by each MLP neural network neuron can be expressed as follows:

$$y = g\left(\sum_{i=1}^{n} w_i \cdot x_i + b\right) \quad (1)$$

where x_i are the neuron inputs, w_i represents the synaptic weight (artificial synapse) belonging to i^{th} input value, θ refers to the activation threshold, u indicates activation potential, y represents the output response produced by the artificial neuron, and $g(.)$ expresses the activation function that must be both differentiable and continuous throughout its domain, which are usually represented by the logistic activation function or by the hyperbolic tangent.

The training process of the neural network consists of applications of ordered steps that are necessary to tune the synaptic weights (w_i) and thresholds (θ) associated to its neurons, with the ultimate goal of generalizing solutions to be produced by its outputs, whose responses are representative of the physical system to which they are mapping. The learning method used here for training MLP network was

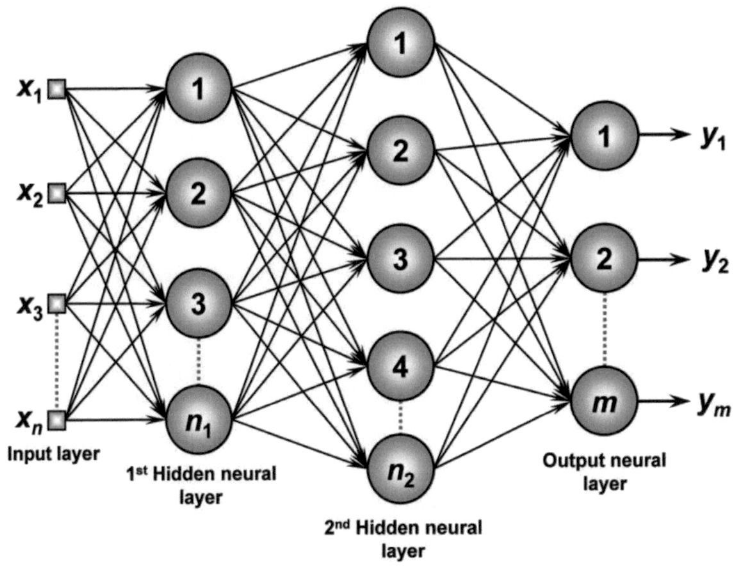

Figure 1.
Illustration of an artificial neural network with multilayer perceptron architecture.

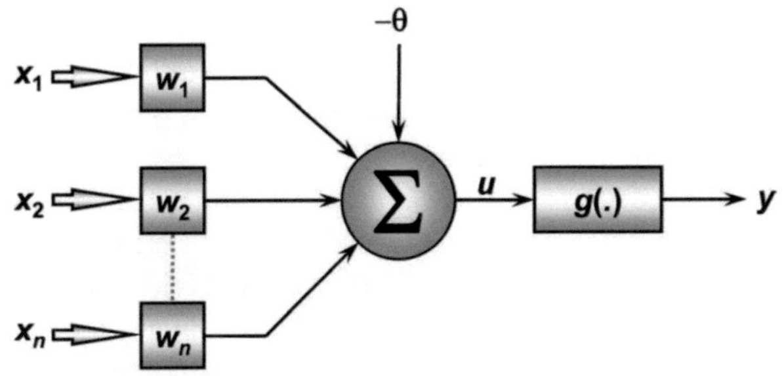

Figure 2.
Schematic diagram of artificial neuron used in MLP neural network.

that based on the Levenberg-Marquardt algorithm, whose detailed steps are presented in [29, 30].

The training of this network was performed in a supervised manner, which consists in having available, considering each sample of the input signals, the respective desired outputs, i.e., each training example is represented by input signals and their respective corresponding outputs. In this case, from a database supplied by electric utilities, the network inputs were the parameters related to the physicochemical tests (**Table 1**) performed on several isolating mineral oil samples, while the desired outputs are the respective values of those gases obtained by chromatographic tests (**Table 2**) using the DGA technique.

Therefore, the MLP network used for this purpose was composed of 10 neurons (first layer) and 20 neurons (second layer), and the activation function adopted for all artificial neurons was the hyperbolic tangent.

4. Experimental results and validation

According to the previously presented sections, several isolating mineral oil samples were also presented to the MLP network in order to select the best architecture for the problem mapping, whose results are reported in the next figures.

Figure 3 shows the estimated levels of gases dissolved in oil. Physicochemical experimental tests data were inserted in the MLP network inputs in order to produce, subsequently, the estimation of all gases dissolved in the insulating mineral oil. Network responses were compared to the values obtained from the chromatographic analyses. It is noteworthy that the MLP network responses are in accordance with the gas concentration.

Based upon **Figure 4**, it is possible to check the MLP network capacity of making adaptation and generalization about the analyzed data, whose results were compared to actual values of H_2 obtained from experimental tests.

Estimation of oxygen (O_2) obtained from 20 samples, which were not included in the training data, is shown in **Figure 5**. Considering this figure, it is possible to verify that the MLP network responses are very near to those obtained from the chromatographic tests.

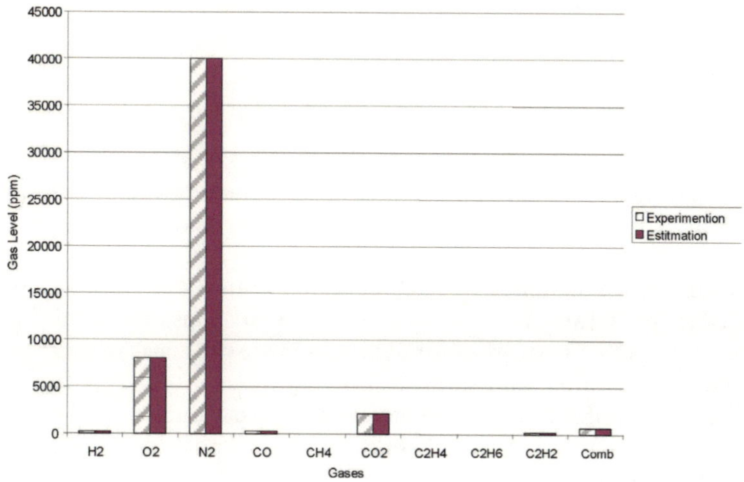

Figure 3.
Estimation of gas levels in the insulating mineral oil.

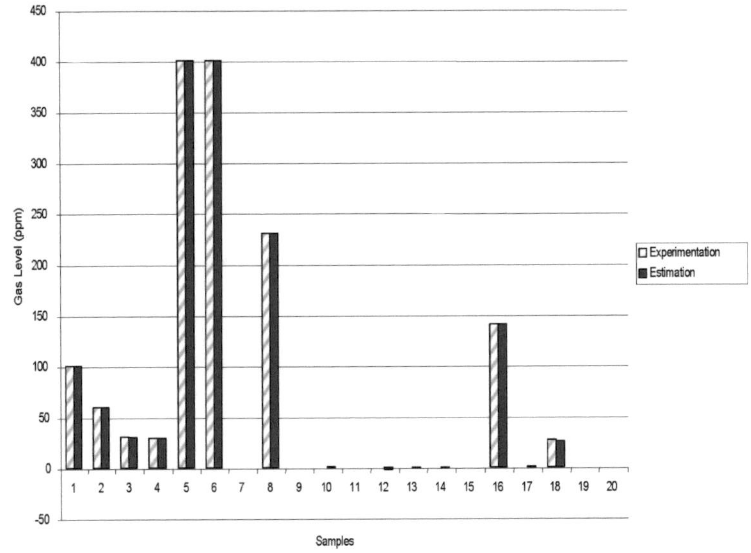

Figure 4.
Estimated levels of H_2 and comparative analysis.

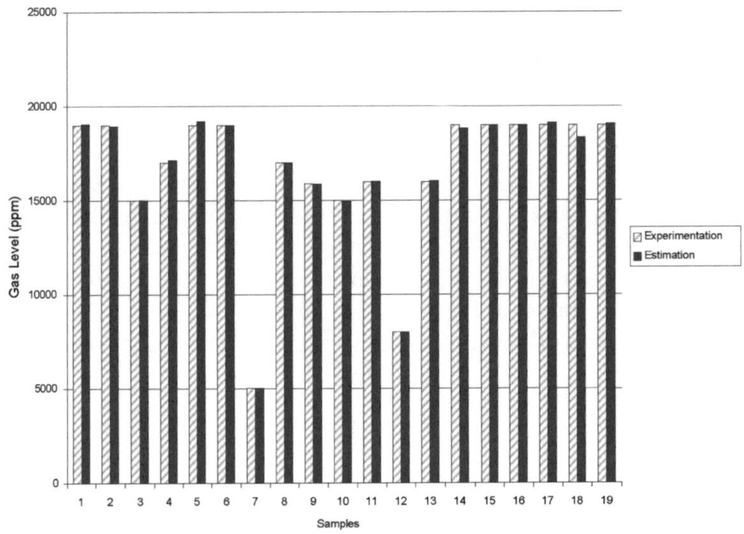

Figure 5.
Estimated levels of O_2 and comparative analysis.

Regarding oxygen (O_2), it is noticed that the MLP network generalization about physicochemical analysis is very reasonable. Therefore, it is appropriate to estimate such parameters.

Estimation of carbon monoxide (CO) obtained from 20 samples, which were not included in the training data, is shown in **Figure 6**. Considering this figure, it is possible to confirm that network computed results are very near to those obtained from the chromatographic analysis.

In addition, it is noted that generalization of CO gas values carried out by the neural network based upon physicochemical analysis is positive. Therefore, it is also appropriate to estimate such gas values.

Estimation of carbon dioxide (CO_2) obtained from 20 samples, which were not included in the training data, is shown in **Figure 7**. Considering this figure, it is

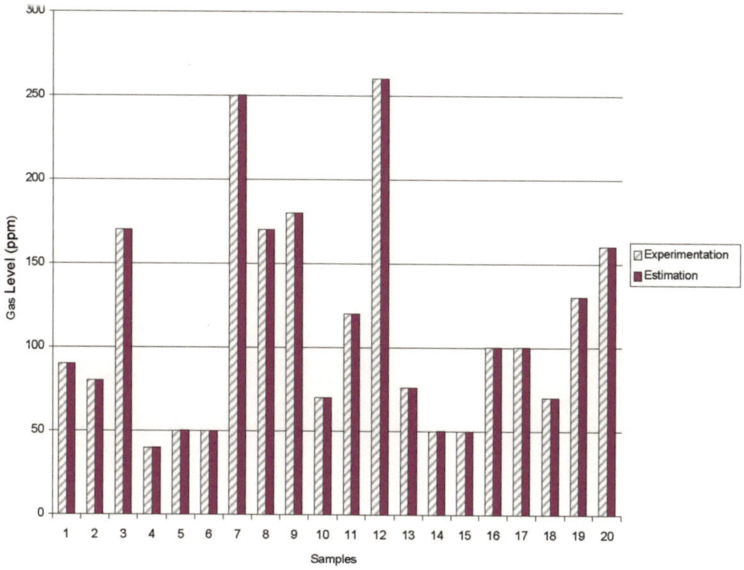

Figure 6.
Estimated levels of CO and comparative analysis.

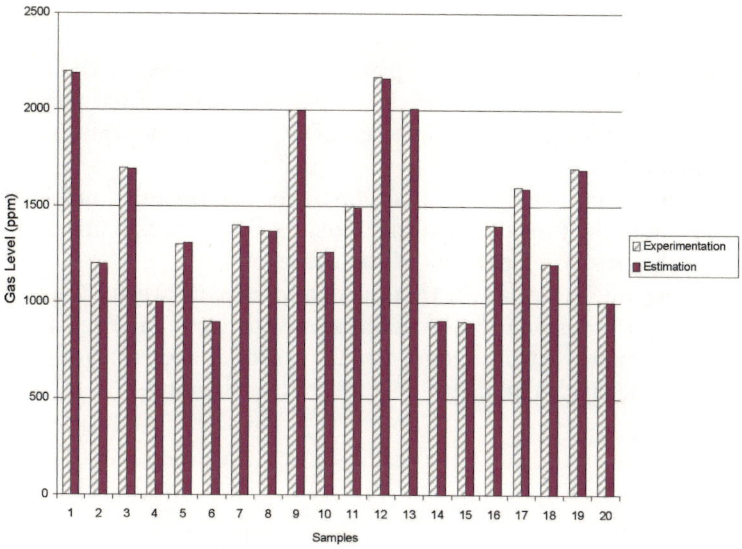

Figure 7.
Estimated levels of CO_2 and comparative analysis.

possible to check that network computed results are also very near to those obtained from the chromatographic analysis.

To summarize, it is verified that the generalization of CO_2 gas values based upon physicochemical analysis is positive. Therefore, it is also appropriate to estimate such parameters.

The following figures illustrate the results provided by the MLP neural network relating the estimated gases in relation to the physicochemical parameters.

Figure 8 illustrates the relationship between CO_2 gas levels with their colors, and these attributes are tabulated from 0.5 (for new oil) to 3.0 (for deteriorated oil). We have here checked that the relationship between CO_2 levels decreases as the color

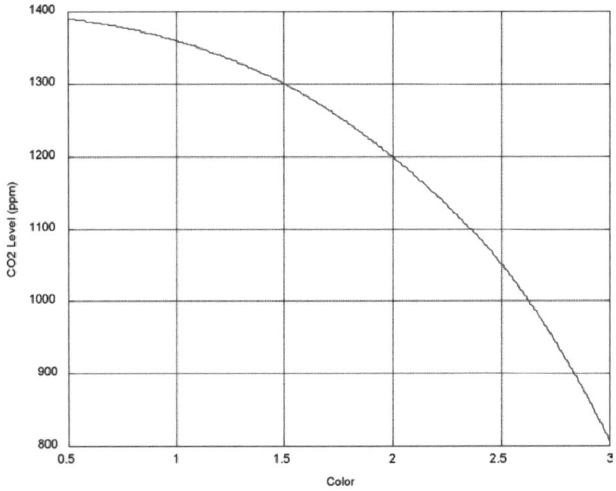

Figure 8.
Relationship between CO_2 gas levels as a function of the oil color.

level increases. This behavior is related to the heating of the insulating mineral oil, which consequently also increases its oxidation level.

On the other hand, as illustrated in **Figure 9**, it is reported that CO_2 gas levels increase linearly as the oil density value also increases. This behavior is related to abrupt changes in temperature over time.

Figure 10 shows that the relationship between CO_2 gas levels and the neutralization index of the insulating mineral oil decreases nonlinearly. This behavior may be related to oil acidity levels, since the increase of CO_2 has direct implications on its neutralization levels.

Figure 11, in contrast to **Figure 9**, has illustrated that the relationship of CO gas levels as a function of insulating mineral oil density decreases linearly, as the behavior of both gases is independent of each other.

Already in **Figure 12**, we can see that the relationship between CO gas levels with their interfacial tension values remains almost constant until tension limit of 25 dyn/cm, and from this critical point, the CO gas levels drop sharply, as the less water present in the oil also implies higher interfacial tension.

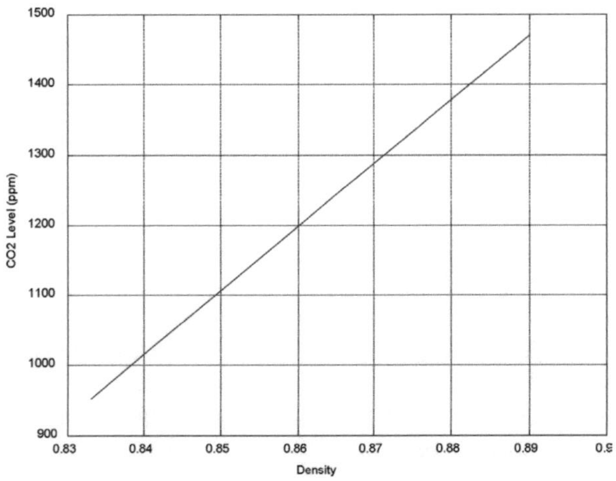

Figure 9.
Relationship between CO_2 gas levels as a function of the oil density.

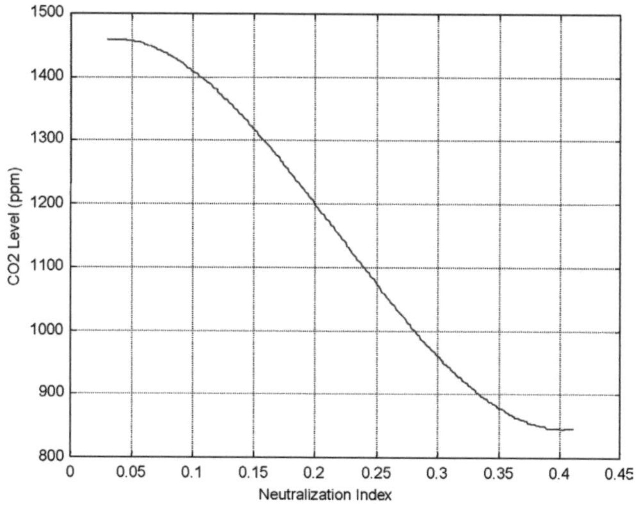

Figure 10.
Relationship between CO_2 gas levels as a function of the oil neutralization index.

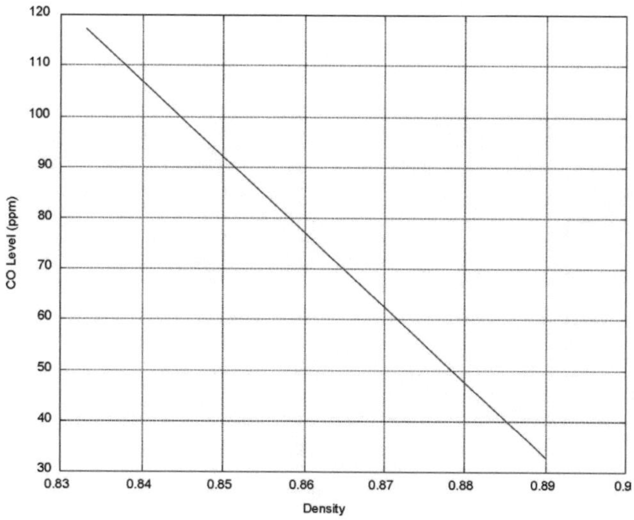

Figure 11.
Relationship between CO gas levels as a function of the oil density.

Figure 13 now shows that CO gas levels as a function of water quantity are also inversely proportional. This behavioral profile is probably associated with temperature, as its decrease also implies a reduction in water evaporation and, consequently, increases its concentration in the insulating mineral oil.

Finally, as shown in **Figure 14**, it is reported that the relationship of O_2 gas levels as a function of oil density has a decreasing profile. Such behavior may be due to changes in acidity levels of insulating oil as density increases.

From the results presented in this section, we demonstrate the possibility of mapping, by means of artificial neural networks, concentration values of gases dissolved in insulating mineral oil from the results of physicochemical tests. To estimate the gases, data from physicochemical tests performed on insulating mineral oil taken from operating transformers were used.

The developed technique allowed the gas estimations in insulating mineral oil and without need to perform the chromatographic tests. Thus, the amount

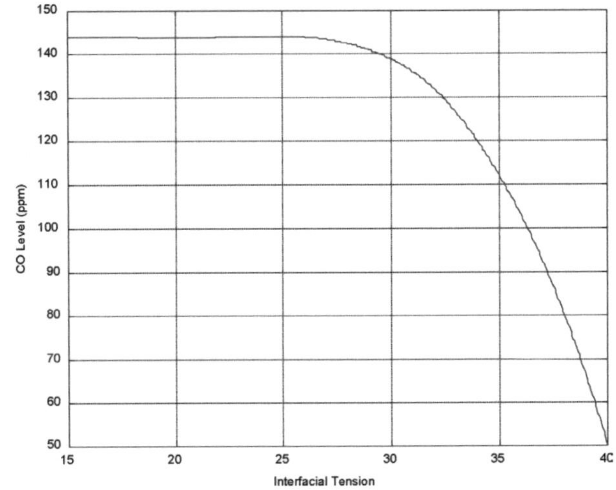

Figure 12.
Relationship between CO gas levels as a function of the oil interfacial tension.

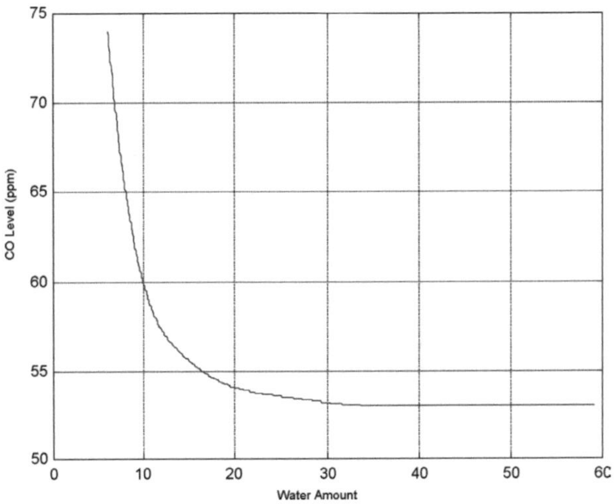

Figure 13.
Relationship between CO gas levels as a function of the water concentration in the insulating mineral oil.

of power supply interruptions can be considerably reduced, and the associated costs in determining the concentration of gases in the mineral oil follow this same downward trend.

5. Conclusion

An approach based on artificial neural networks was presented in this article, and its target was to estimate gas levels dissolved in insulating oil. Based upon data obtained from physicochemical analyses, decisions for transformer maintenance may be supported by the network; it can even indicate whether chromatographic analyses are necessary.

The artificial neural network approach proved to be efficient for estimating the gas levels dissolved in insulating oil. Based upon input data provided by

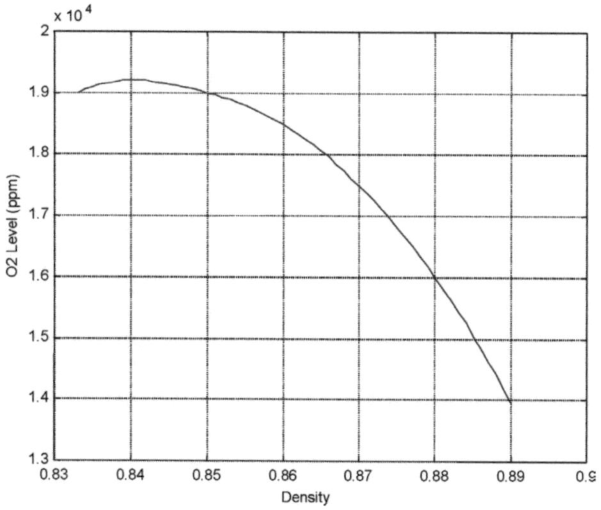

Figure 14.
Relationship between O_2 gas levels as a function of the oil density.

physicochemical analyses, estimation of any gas obtained from chromatographic analyses becomes possible (without physicochemical analysis). Furthermore, artificial neural network approach was developed in order to set up the relationship between attributes which did not have any apparent relationship.

Finally, the developed neural approach is innovative, and it is also the first one that performs the estimation of dissolved gases in insulating mineral oil from the results of physicochemical tests. Therefore, in order to authenticate the developed methodology, the results provided by the artificial neural network were compared with those actual values obtained by chromatographic tests.

Acknowledgements

The authors thank the ANEEL Research and Development Program under contract number PD-0068-0037/2016.

Author details

Ivan Nunes da Silva[1]*, Rogério Andrade Flauzino[1], Danilo Hernane Spatti[1], Renato Pagotto Bossolan[2] and Bruno Augusto Trevisam[2]

1 University of São Paulo, São Carlos, SP, Brazil

2 São Paulo State Electric Power Transmission Company, ISA/CTEEP, São Paulo, SP, Brazil

*Address all correspondence to: insilva@sc.usp.br

References

[1] Ferreira LG. Procedures for regeneration of oils insulating minerals. Modern Electricity. 1996;**25**:39-49

[2] Liang Z, Parlikad A. A Markovian model for power transformer maintenance. International Journal of Electrical Power & Energy Systems. 2018;**99**:175-182

[3] Koksal A, Ozdemir A. Improved transformer maintenance plan for reliability centred asset management of power transmission system. IET Generation, Transmission and Distribution. 2016;**10**(8):1976-1983

[4] Wattakapaiboon W, Pattanadech N. The state of the art for dissolved gas analysis based on interpretation techniques. In: International Conference on Condition Monitoring and Diagnosis (CMD), Xi'an, China; 2016. pp. 60-63

[5] Cox R. Categorizing transformer faults via dissolved gas analysis. In: 19th IEEE International Conference on Dielectric Liquids, Manchester, UK; 2017. pp. 1-3

[6] Ariffin MM, Ishak MT, Hamid MHA, Katim NIA, Ishak AM, Azis N. Ageing effect of vegetable oils impregnated paper in transformer application. In: International Conference on High Voltage Engineering and Power Systems, Sanur, Indonesia; 2017. pp. 183-187

[7] Ariffin MM, Ishak MT, Hamid MHA, Katim NIA, Ishak AM, Azis N. Comparative studies of the stability of various fluids under electrical discharge and thermal stresses. IEEE Transactions on Dielectrics and Electrical Insulation. 2015;**22**(5):2491-2499

[8] Silva IN, Spatti DH, Flauzino RA, Liboni LHB, Alves SFR. Artificial Neural Networks – A Practical Course. 1st ed. Zurich, Switzerland: Springer; 2017

[9] Khuntia SR, Rueda JL, Bouwman S, Van der Meijden MAMM. A literature survey on asset management in electrical power [transmission and distribution] system. International Transactions on Electrical Energy Systems. 2016;**26**:2123-2133

[10] Zhong J, Li W, Wang C, Yu J, Xu R. Determining optimal inspection intervals in maintenance considering equipment aging failures. IEEE Transactions on Power Systems. 2017;**32**:1474-1482

[11] Liu C, Huang G, Zhang K, Wen F, Salam MA, Ang SP. Asset management in power systems. In: 10th International Conference on Advances in Power System Control, Operation & Management, Hong Kong, China; 2015. pp. 1-5

[12] Shah SWA, Mahmood MN, Das N. Strategic asset management framework for the improvement of large scale PV power plants in Australia. In: Australasian Universities Power Engineering Conference, Brisbane, Australia; 2016. pp. 1-5

[13] ABNT–NBR–11343. Petroleum products - Determination of aniline point and mixed aniline point. Brazilian Association of Technical Standards; 2003

[14] ABNT–NBR–7070. Sampling of gases and mineral insulating oil of electrical equipments and free and solved gases analysis. Brazilian Association of Technical Standards; 2006

[15] Haykin S. Neural Networks and Learning Machines. 3rd Ed. Upper Saddle River, USA: Prentice Hall; 2008

[16] ABNT–NBR–10441. Petroleum products – Transparent and opaque liquids – Determination of kinematic

viscosity and calculation of dynamic viscosity. Brazilian Association of Technical Standards; 2007

[17] ASTM–D1500. Standard test method for ASTM color of petroleum products (ASTM color scale). American Society for Testing and Materials; 2017

[18] ABNT–NBR–7148. Petroleum and petroleum products – Determination of density, relative density and °API – Hydrometer method. Brazilian Association of Technical Standards; 2013

[19] IEC–156. Insulating liquids – Determination of the breakdown voltage at power frequency – Test method. International Electrotechnical Commission; 1995

[20] ABNT–NBR–6869. Electric insulating liquids – Determination of the dielectric breakdown voltage (disk electrodes). Brazilian Association of Technical Standards; 1989

[21] IEC–247. Measurement of relative permittivity, dielectric dissipation factor and d.c. resistivity of insulating liquids. International Electrotechnical Commission; 1979

[22] ASTM–D974. Standard test method for acid and base number by color-indicator titration. American Society for Testing and Materials; 2014

[23] ABNT–NBR–6234. Mineral insulating oil – Determination of interfacial tension of oil-water – Test method. Brazilian Association of Technical Standards; 2015

[24] ABNT–NBR–10504. Mineral insulating oil – Determination of oxidation stability – Test method. Brazilian Association of Technical Standards; 1988

[25] ABNT–NBR–11341. Petroleum products – Determination of the flash and fire points by Cleveland open cup. Brazilian Association of Technical Standards; 2004

[26] ABNT–NBR–8148. Measurement of the average viscosimetric degree of polymerization of new and aged electrical papers and paperboards. Brazilian Association of Technical Standards; 2000

[27] ABNT–NBR–5755. Determination of water in insulating liquids (Method of Karl Fischer). Brazilian Association of Technical Standards; 1984

[28] Milasch M. Maintenance of Transformers in Insulating Liquid. Ed. São Paulo, Brazil: Edgar Blucher; 1984

[29] Hagan MT, Menhaj MB. Training feedforward networks with the Marquardt algorithm. IEEE Transactions on Neural Networks. 1994;**5**(6):989-993

[30] Shinde SB, Sayyad SS. Cost sensitive improved Levenberg Marquardt algorithm for imbalanced data. In: IEEE International Conference on Computational Intelligence and Computing Research, Chennai, India; 2017. pp. 1-4

Chapter 7

Efficient Asset Management Practices for Power Systems Using Expert Systems

Danilo Spatti, Luisa H.B. Liboni, Marcel Araújo, Renato Bossolan and Bruno Vitti

Abstract

Electric power companies have high financial costs due to poor asset management practices. Therefore, it is crucial to use decision-making processes to decrease the global costs of an active asset and to extend its lifetime to a maximum. Asset management programs, which are frequently used to tackle optimization problems, aim to guide the use of the physical assets of a business, mainly by optimizing their lifetime. Efficient asset management practices establish operation and maintenance for each equipment, from the time the equipment is acquired until the appropriate time for its replacement. So, based on these assumptions, we propose a method to assist asset management decision-making in the electric power companies, which is embodied by computer software.

Keywords: expert systems, asset management, power systems

1. Introduction

The big challenge involving asset management in electric systems is to seek a solution that enables the electric sector to reconcile interests with the environmental, economic, market, technological, regulatory and corporate constraints.

Still, the first parameters to be considered in the asset management of electrical systems can be listed as follows:

- Equipment lifetime
- Operating conditions
- External conditions

Additionally, directly or indirectly, operational context aspects must also be considered, such as:

- Type of operation requests (existence of redundancies or equipment in standby)
- Requirement level to be met or minimum performance requirement
- Operational safety risks to be assumed

- Environmental standards
- Equipment life cycle
- Maintenance logistics
- Regulatory standards and legislation

With a large amount of information currently available from assets, it is critical that standalone or semi-autonomous machine learning techniques should be able to extract knowledge and increase the performance or robustness of a system or process. These algorithms and techniques are vast and are also used in non-database applications—for example; such methods can directly interact with the environment to accomplish the tasks discussed above [1, 2].

The computational tools and techniques based on machine learning help to solve problems related to the extraction of information and knowledge from data. For example, machine learning applied to data mining is capable of performing specific tasks such as:

- Pattern recognition
- Data processing
- Data clustering
- Feature selection (selection of the most critical variables in a process, system, or database)
- Regressions
- Data sorting

The purpose of this chapter is to present a case study involving the processing of asset databases of an electrical transmission system in order to create subsidies for the development of an efficient asset management system.

2. Database studies

The databases in this case study are composed of data from power transformers. Such databases contain nominal values from the devices, data from laboratory tests, and maintenance data. The database consists of 6929 records, with the attributes shown in **Table 1**.

Data preprocessing methods consists of database conditioning, inconsistent data correction and the analysis of the data through relational graphs and probability distributions in order to extract knowledge and useful information.

Figure 1 shows a histogram of the transformers with respect to their manufacturing date and **Figure 2** shows a histogram of the transformers with respect to their age. **Table 2** shows the records for the voltage attribute of the transformers.

By using data from maintenance records, one can classify maintenance tasks as Preventive Maintenance and Corrective Maintenance. One of the critical analysis

Manufacturer	Year of construction
Time in operation	Feeder phase
Did achieve its lifetime according to regulation policies?	Power
Location	Voltage
Level of priority of the maintenances	Preventive maintenance dates
Technical maintenance team	Corrective maintenance dates
Age of the transformer when corrective maintenance was performed	

Table 1.
Attributes from the power transformers database.

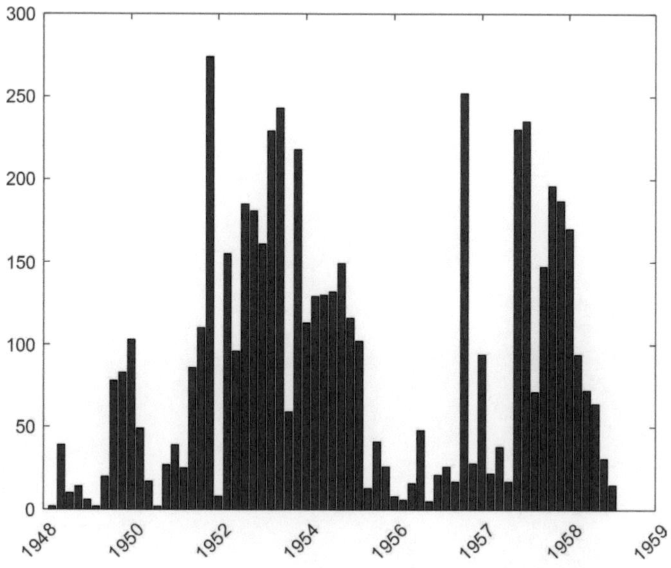

Figure 1.
Histogram (count percentage) of the transformers with respect to their manufacturing date.

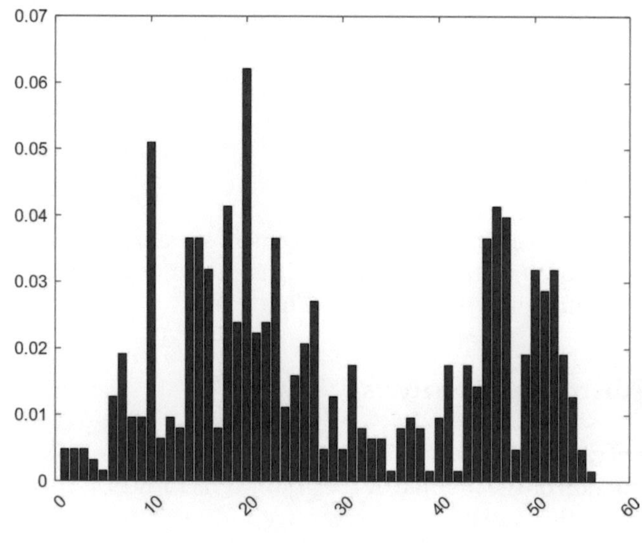

Figure 2.
Histogram (count percentage) of the transformers with respect to their age.

Voltage (kV)	Records
13.8	1221
34.0	194
69.0	95
88.0	1497
138.0	2092
145.0	4
230.0	545
345.0	451
440.0	690

Table 2.
Voltage attribute from the power transformers database.

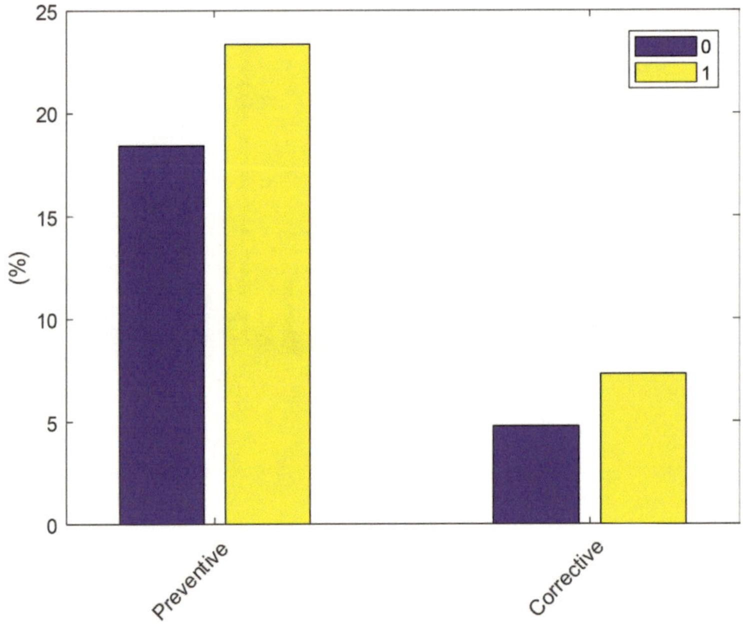

Figure 3.
Histogram (count percentage) of the transformers with respect to preventive and corrective maintenance, taking into account if the lifetime was achieved (1) or not (0).

that must be done by an efficient asset management system is to take into account the lifetime of the power transformer and the maintenance procedures held during its lifetime [3, 4]. Such analysis is shown in **Figure 3**.

3. Statistical distribution analysis

One can verify, in **Figure 4**, the relationship between the probability of maintenance as a function of the lifetime of the power transformers. In this figure, the cumulative distribution functions of preventive and corrective maintenance are shown and have, clearly, a bathtub shape (**Figure 5**) [5].

The curves in **Figure 4** raise a management question, and, therefore, explain the importance of such analysis: the more preventive maintenance, the less corrective

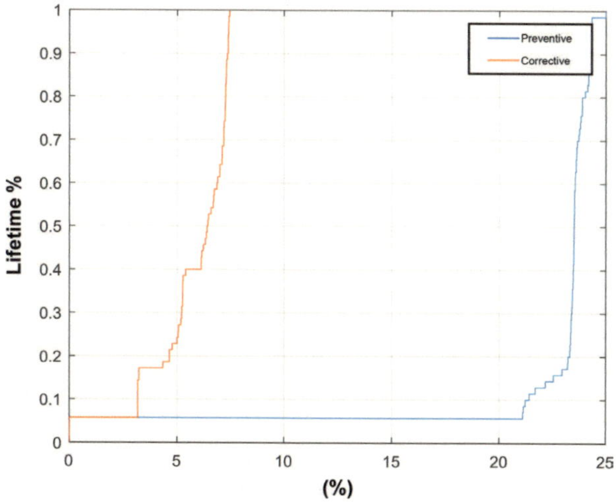

Figure 4.
Cumulative probability function of the power transformers.

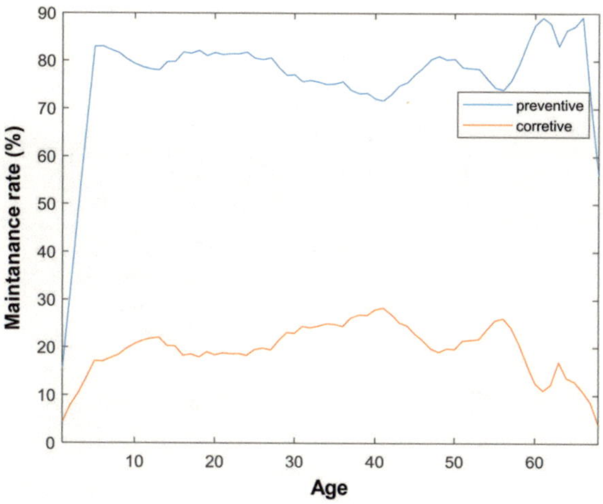

Figure 5.
Maintenance rate with respect to the age of the assets.

maintenance? This hypothesis can be tested by a correlation analysis, which indicates a Correlation between preventive and corrective maintenance of −0.9367, i.e., the two types of maintenances are firmly negatively correlated.

One can observe that preventive maintenance in equipment such as power transformers is made in a more uniform way throughout its operating lifespan. This result corroborates with the fact that this kind of equipment is crucial and should be robust and faultless, as is shown in **Figure 6**.

To explore the hypothesis mentioned above, two manufacturers were chosen, namely A and B, and the following analysis was performed. The results are shown in **Figures** 7 and **8**.

By analyzing these curves, some questions can be raised concerning asset management in order to help decision-making. For example, in this scenario, would it be possible to purchase more equipment with this specific nominal voltage from Manufacturer 2?

These analyses can be extended for other voltage classes, such as 345 and 440 kV, as can be seen in **Figures 9** and **10**.

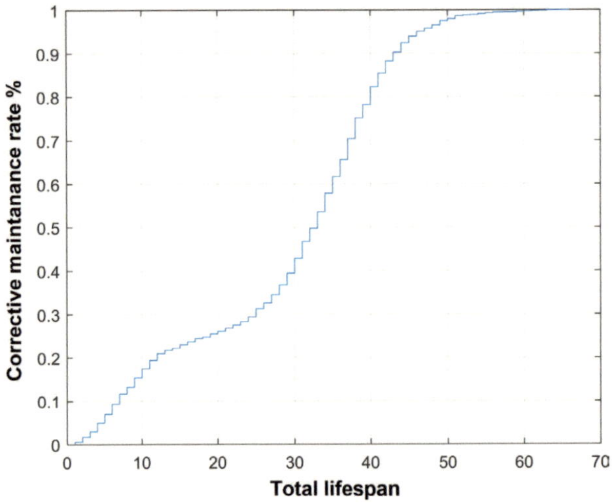

Figure 6.
Cumulative probability function for corrective maintenance.

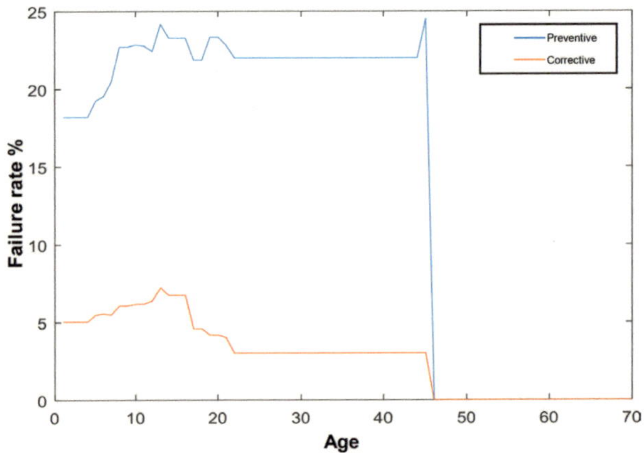

Figure 7.
Maintenance rate with respect to the age of the assets for manufacturer A.

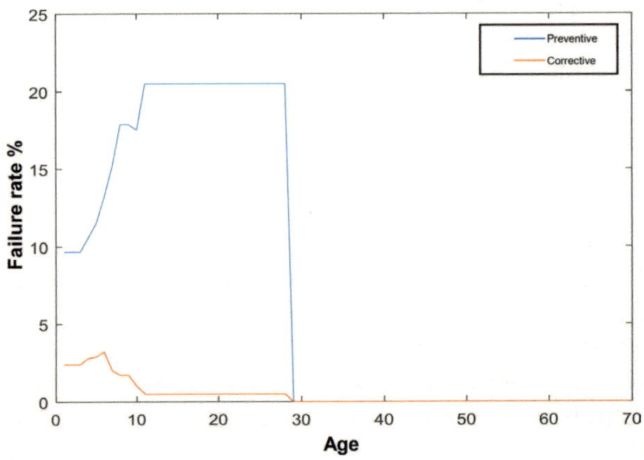

Figure 8.
Maintenance rate with respect to the age of the assets for manufacturer B.

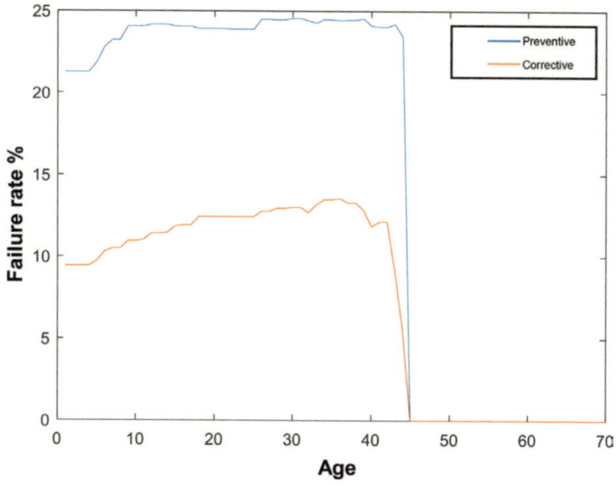

Figure 9.
Maintenance rate with respect to the age of the assets for voltage 345 kV.

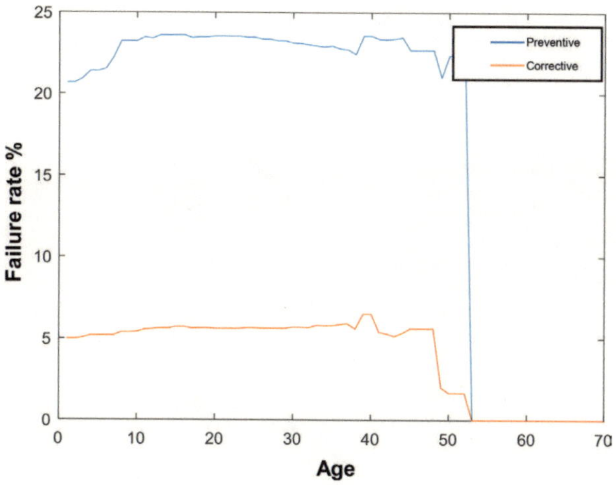

Figure 10.
Maintenance rate with respect to the age of the assets for voltage 440 kV.

4. Risk analysis

Defining a process for creating a condition index for an asset is very important since the index can be used for asset management. For example, managers could consider alternative manufacturer scenarios, replacement or new acquisitions. Thus, the condition index of equipment can be an important feedback tool for the electric utility regulator since good maintenance practices dictated by the regulator, such as the frequencies for maintenance, can be updated [6, 7].

Also, risk assessment allows the analysis of failure probability, as shown in **Figures 11** and **12**, in which all power transformers in the database were analyzed.

Alternatively, it can also be established that the reliability function R (t) is given by R (t) = 1 − F (t), where F (t) is the cumulative probability function. Function R (t) can be used to analyze the probability of failure of an asset above a particular range or age, as shown in **Figure 13**.

As a case study, one can then investigate what the failure rate would be for a particular transformer cluster. Now, suppose the cluster composed of all equipment

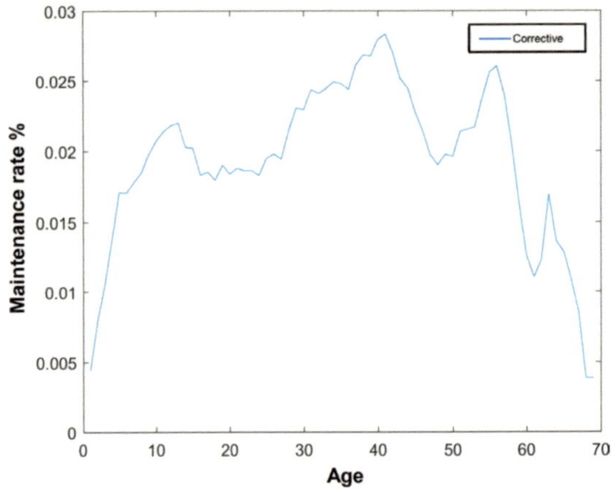

Figure 11.
Corrective maintenance rate against operating time.

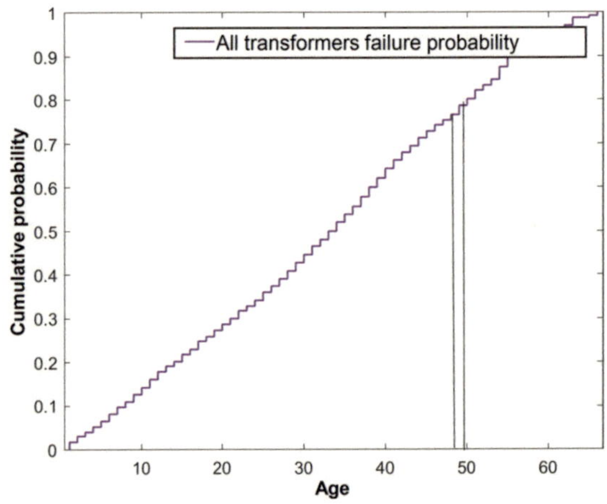

Figure 12.
Probability of failure of all transformers with respect to their age.

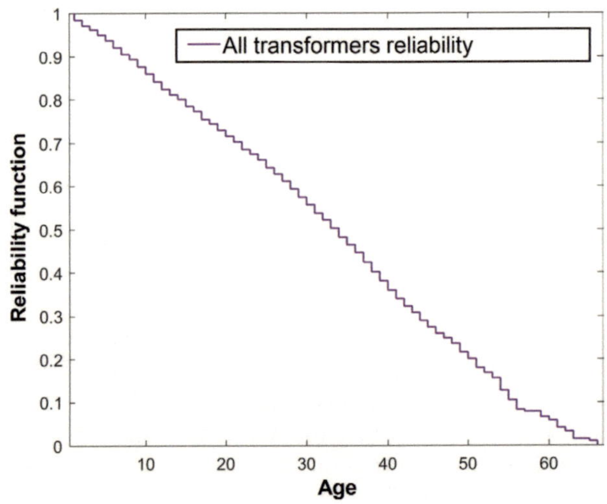

Figure 13.
Reliability rate of all transformers with respect to their age.

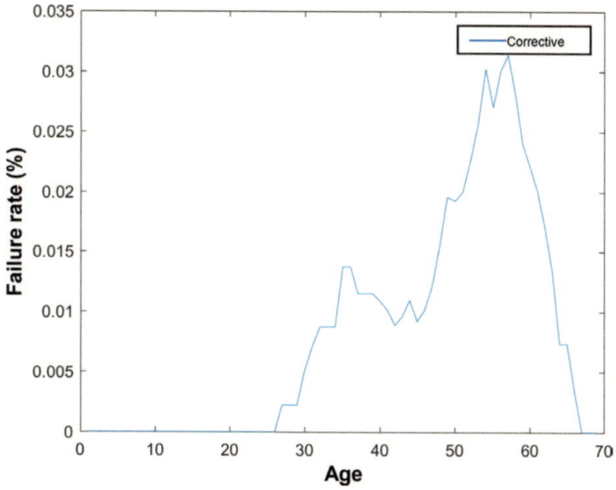

Figure 14.
Manufacturer 1 maintenance rate (138 kV) with respect to operating time.

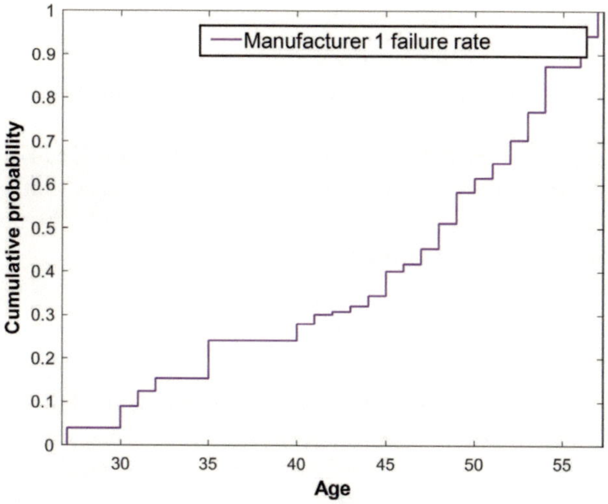

Figure 15.
Failure rate of manufacturer 1 transformers (138 kV) with respect to their age.

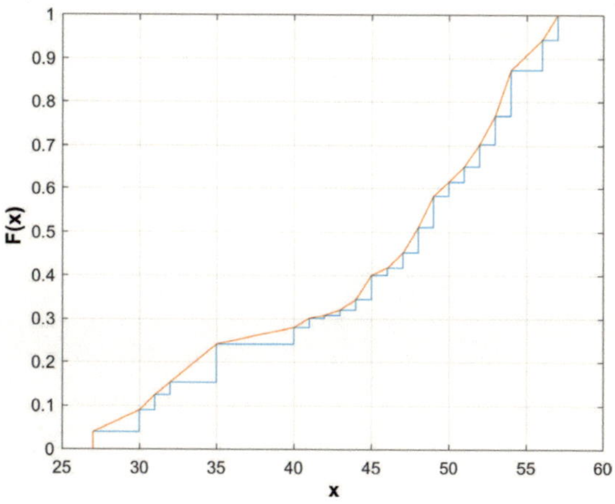

Figure 16.
Empirical distribution approximation.

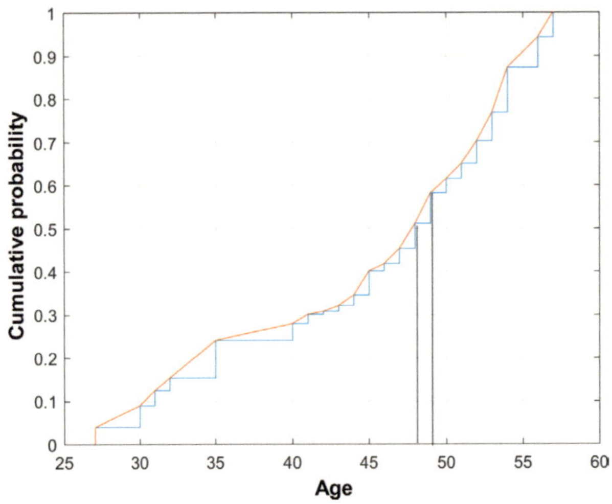

Figure 17.
Transformer ID 147511 (manufacturer 1) failure rate will increase by 7% between 48 and 49 years of age of the asset.

from Manufacturer 1 and with nominal voltage of 138 kV. We then obtain the results shown in **Figures 14** and **15. Figure 16** shows the approximation of this distribution.

In this case, for a particular transformer, i.e., transformer ID 147511 (manufacturer 1 // voltage class 138 kV // age 48 years), one can observe that the failure rate of this asset will increase 7%, between 48 and 49 years of age. The results of such analysis can be seen from the examination of **Figure 17**.

5. Critical state index estimation

The so-called critical state index allows one to verify if the failure rate of the asset is coherent with the moment in its life cycle. Thus, it becomes possible, by considering the age of the asset and its history of failures, to point out the assets that require distinguished attention [7].

There are different methodologies for identifying and determining the type of defect or failure in equipment immersed in insulating mineral oil based on the results of chromatographic analysis. These methods, such as those listed as follows, are directives for estimating a state index for transformers:

- Key Gas
- Roger
- Doernenburg
- IEC ratio
- IEEE
- Duval triangle

These methods analyze gas concentration limits to estimate the state of the transformer, i.e., gas concentrations are used to diagnose the equipment. To make

Gas	Limit L1 (ppm)	Limit G2 (ppm per month)
H_2	100	50
CH_4	75	38
C_2H_2	3	3
C_2H_4	75	38
C_2H_6	75	38
CO	700	350
CO_2	7000	3500

Table 3.
Duval criteria for identifying transformer defects.

this diagnosis, that is, to quantify the probability of a transformer defect, the following information in **Table 3** can be employed.

To confirm the existence of a fault, one of the gases shown in **Table 3** must have a concentration equal to or higher than L1 and more significant than the value indicated by G2. Thus, this methodology has an appropriate way to evaluate the condition of a transformer based on the historical data about its chromatographic assays.

When the historical data are not available, the IEEE method can be more efficient, since it uses four conditions to identify failure states:

- Condition 1 refers to a normal operating condition, that is, if the dissolved gases are at levels below those presented in **Table 3**, then the transformer is considered to be operating correctly.

- Condition 2 indicates a possible failure and many mineral oil samples should be taken to determine the tendency of gas growth.

- Condition 3 indicates a high level of cellulose and/or mineral oil decomposition, and there is probably a transformer failure.

- In condition 4 there is an indication of excessive cellulose and/or mineral oil decomposition. Continued operation under Condition 4 may result in a total fault.

The joint assessment between the equipment critical state index and its risk analysis, which is based on the history of maintenance procedures, allows for a more assertive decision-making process. To illustrate how the decision-making process can be enriched with both information, consider a 440 kV power transformer that entered into operation in 1981. The graph shown by **Figure 18** illustrates the number of accumulated failures since the last total fault, which has required intervention, as a function of the operating time.

Figure 18 shows that a number of failures between 1.3 and 1.6 failures were expected since the last intervention. This analysis is made with a 90% confidence level by considering the asset history and the database of similar assets.

In order to construct curves involving the operating state of the devices, more than 25,000 gas chromatography assays were evaluated. The tests date from 1977 to the present day, and each test was evaluated according to the standards: IEEE, IEC, Duval, Doernenburg, Roger, and Key Gas. The IEEE criterion was adopted to evaluate the operating condition, as represented by **Figures 19–22**.

The integrated Analysis, which is made by considering the mean behavior from all assays can be shown in **Figure 23**.

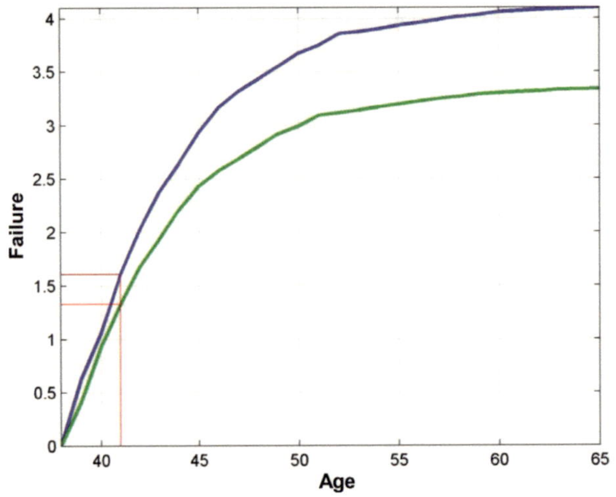

Figure 18.
Number of failures since the last intervention for a given transformer.

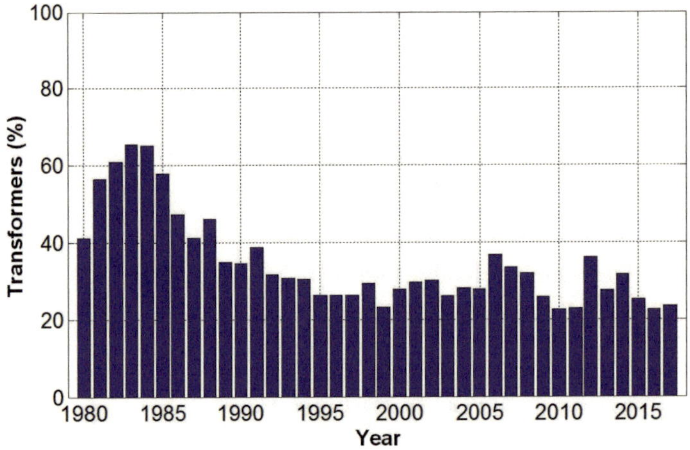

Figure 19.
Percentage of transformers in condition 1.

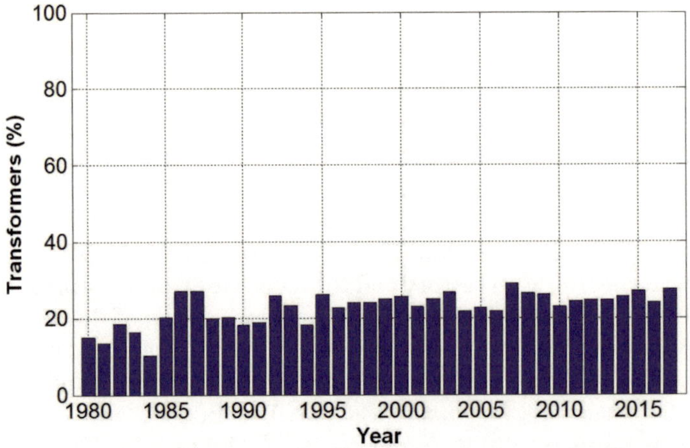

Figure 20.
Percentage of transformers in condition 2.

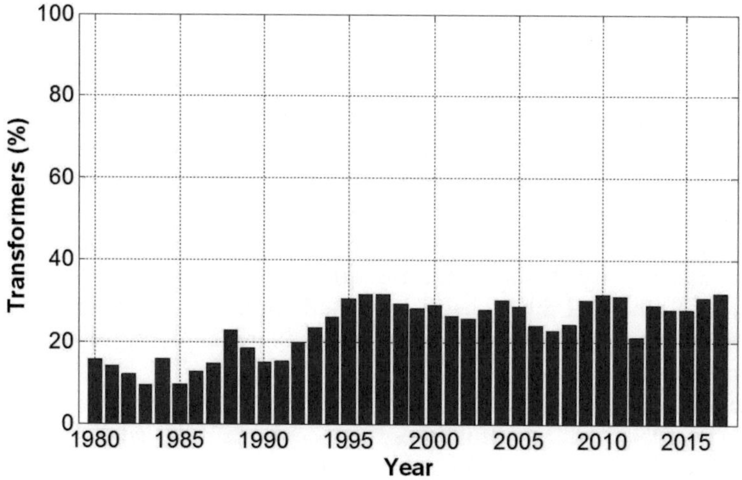

Figure 21.
Percentage of transformers in condition 3.

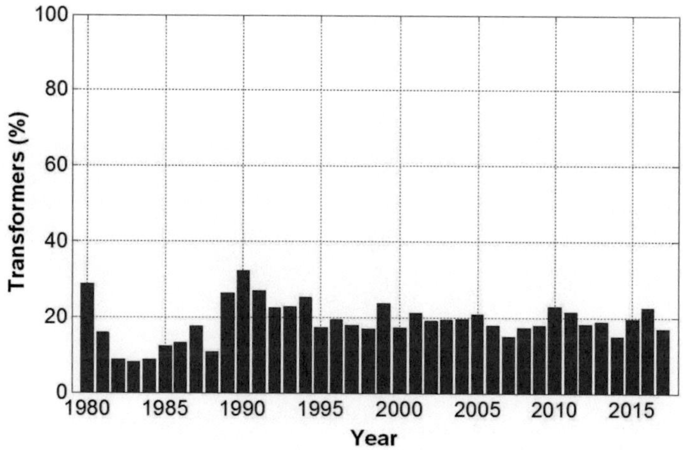

Figure 22.
Percentage of transformers in condition 4.

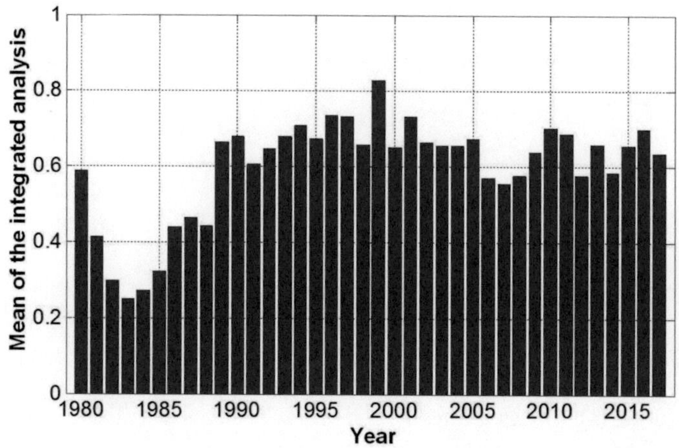

Figure 23.
Integrated analysis.

6. Conclusions

Critical assets in the transmission and distribution industries need special care and attention, mainly regarding the aging of the equipment since their lifespan impacts profits as well as the reliability and safety of the electric system.

As the lifetime of these devices become longer, it is justifiable to develop methods to identify their health condition, taking into account not only historical data but also all available asset management tools that companies currently own.

As the vast majority of companies still struggle to learn from the abundant data acquired, such a method should have significant implications in helping managers evaluate and question in-company policies regarding manufacturers and preventive maintenance practices.

In this context, the methodology presented in this chapter can be applied to determine the lifespan of transmission system equipment, based on the determination of failure rates, by using statistical analysis.

Acknowledgements

The authors thank the ANEEL R & D Program, contract number PD-0068-0037/2016.

Author details

Danilo Spatti[1*], Luisa H.B. Liboni[2], Marcel Araújo[3], Renato Bossolan[4] and Bruno Vitti[4]

1 University of São Paulo, São Carlos, Brazil

2 Federal Institute of Education, Science, and Technology of São Paulo, Sertãozinho, Brazil

3 Federal Rural University of Pernambuco, Recife, Brazil

4 São Paulo State Electric Power Transmission Company, São Paulo, Brazil

*Address all correspondence to: spatti@icmc.usp.br

References

[1] Khuntia SR, Rueda JL, Bouwmanb S, Meijden M. Classification, domains and risk assessment in asset management: A literature study. In: 2015 50th International Universities Power Engineering Conference (UPEC). 2015

[2] Nowlan FS, Heap HF. Reliability Centered Maintenance. National Technical Information Service, EUA, Report no. AD/A066-579; 1978

[3] Lapworth JA, Wilson A. The asset health review for managing reliability risks associated with ongoing use of ageing system power transformers. In: 2008 IEEE CMD–International Conference on Condition Monitoring and Diagnosis. 2008. pp. 605-608

[4] Jahromi A, Piercy R, Cress S, Service J, Fan W. An approach to power transformer asset management using health index. IEEE Electrical Insulation Magazine. 2009;**25**(2):20-34

[5] Carneiro JC, Jardini JA, Brittes JLP. Substation power transformer risk management: Reflecting on reliability centered maintenance and monitoring. In: 2012 IEEE/PES T&D–LA–Sixth Transmission and Distribution Conference and Exposition Latin America. 2012. p. 8

[6] Nemeth B, Benyo T, Jager A, Csepes G, Woynarovich G. Complex diagnostic methods for lifetime extension of power transformers. In: 2008 IEEE ISEI–International Symposium on Electrical Insulation. 2008. pp. 132-135

[7] Zhang X, Gockenbach E. Asset-management of transformers based on condition monitoring and standard diagnosis. IEEE Electrical Insulation Magazine. 2008;**24**(4):26-40

Chapter 8

Computational Intelligence to Estimate Fault Rates in Power Transformers

Danilo Spatti, Luisa H.B. Liboni, Marcel Araújo, Renato Bossolan and Bruno Vitti

Abstract

Asset management in power transmission systems is one of the significant practices carried out by power companies. With the aging of the devices, the development of optimized tools, capable of considering failure rates, regulatory scenarios, and operational parameters, is increasingly mandatory. The purpose of this work is to present a statistics-based tool for optimized asset management. For such an objective, we have developed a computational method based on database processing and statistical studies that can support decision-making on preventive maintenance in the equipment of the electric sector. The final system interface is Business Intelligence-based.

Keywords: computational intelligence, failure rates, power transformers

1. Introduction

In this chapter, we will describe a software, based on Business Intelligence, that has health data (test data, inspections, and operation) of assets as inputs, as well as their technical and constructive characteristics [1, 2]. The software has as outputs the possible categorization of assets into families, the calculation and analysis of failure rates, and the detection of current or incipient anomalies. Therefore, the software includes relational graphs, statistical analysis, and machine learning tools.

In **Figure 1**, the central conception of the system, which is named as the ISA CTEEP Asset Management Support System (AMSS), is shown. Thus, data from critical assets, such as power transformers, will also be considered as inputs to the software. In addition to making trend and relational graphs, the output module for anomaly identification and failure rate calculation, by analyzing test and inspection data, indicates the probability of a particular asset to needing special care. The software architecture was developed so that both input and output modules are easily accessible to the users through a graphical interface [3].

The algorithms that perform the relational and critical analysis of the health of an asset and calculate its failure rate are hosted within the program called Analysis Engine.

Input data may be made available manually by the user or may be directly acquired at the ISA CTEEP database. The software should query the database

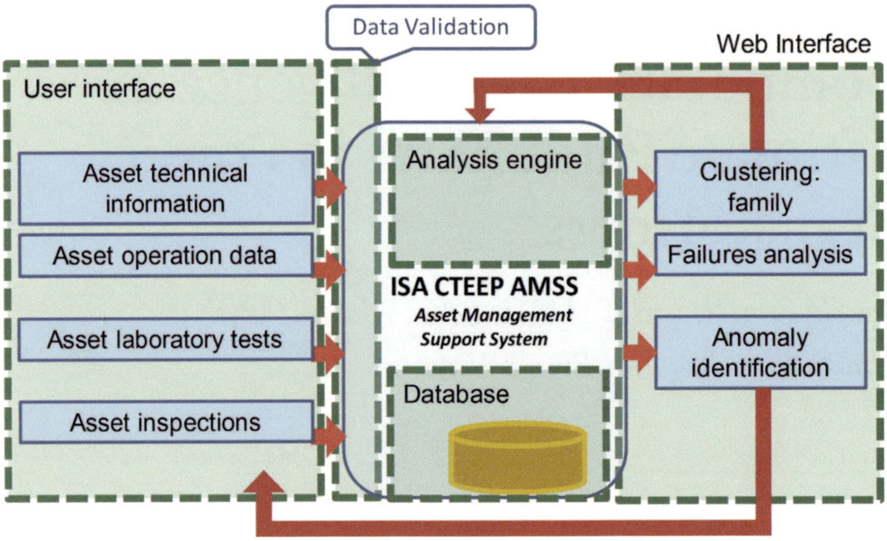

Figure 1.
Structure of the AMSS system.

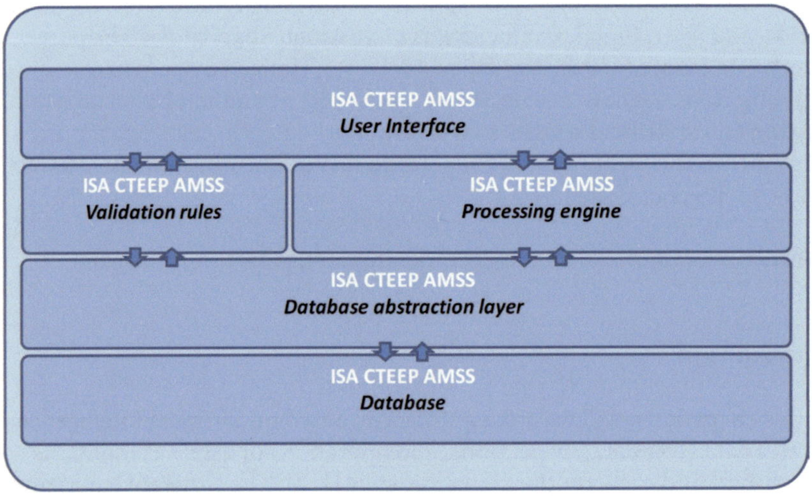

Figure 2.
Layers of the AMSS system.

through a database abstraction (interpreter) layer. These architectural details can be seen in **Figure 1**. **Figure 2** shows, in detail, the functional layers of the software.

The assessment of the health of major interest assets, such as transformers, autotransformers, and reactors, will be made considering three aspects [4]:

a. History of operation and maintenance

b. Routine tests:

- Insulating mineral oil tests:
  - Physicochemical tests
  - Gas chromatography

- ○ Liquid chromatography
- ○ Transformer power factor test

c. Specialized tests:

- Insulation resistance
- Short circuit impedance
- Ohmic resistance
- Transformation ratio
- Inspection of accessories
- Special tests
- Etc.

2. Modeling fault as states

Maintenance procedures are considered inputs to the fault analysis we shall present. In total, there are 5371 maintenance records available in the database from the end of 2008 to the present moment. Maintenance records are mined in order to find relevant information for the analysis. Thus, nonrelevant corrective maintenance records are excluded from the analysis. **Figure 3** shows the relationship between the number of maintenance records as a function of the year of occurrence and the age of the asset [5].

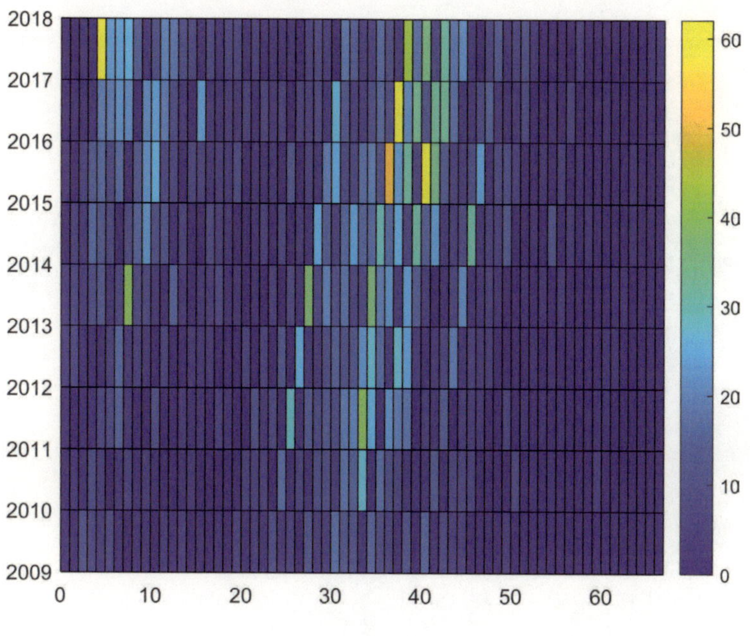

Figure 3.
Summary of the number of records according to the year of occurrence and the age of the asset.

From **Figure 3** it is possible to observe the existence of sparse data, as better highlighted in **Figure 4**. The same sparsity can be perceived in **Figure 5**, where the relationship between the number of equipment and its age is shown.

Assets can be categorized into states. The ideal state in which the equipment must operate is the **NORMAL** state.

Preventive Maintenance processes can ensure that the asset remains in its **NORMAL** state.

However, even with preventive maintenance procedures, there is a transition from the **NORMAL** state to the **DEFECT** state, as shown in **Figure 6**. The state transition occurs given a defect rate.

The transition from the **DEFECT** state (or fault state) to the **NORMAL** state occurs when **Corrective maintenance** actions are made. The transition takes place through a Corrective Maintenance rate, as represented in **Figure 7**.

Still, there is the **FAILURE** state, which is characterized by the complete withdrawal of operation of the asset. The transition takes place through a failure rate, as in **Figure 8**.

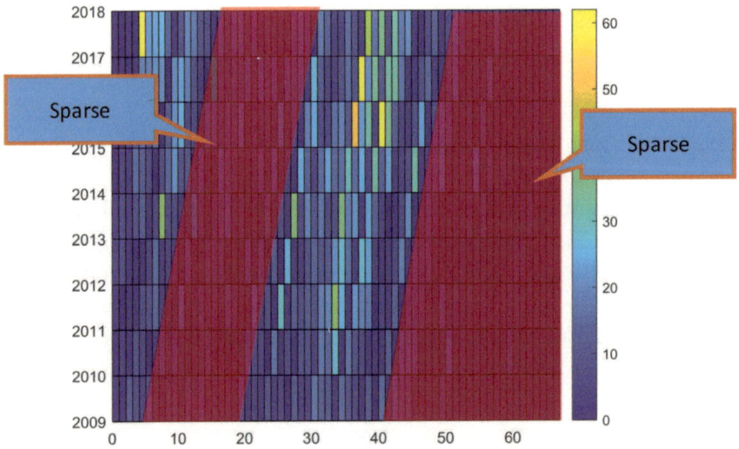

Figure 4.
Number of maintenance records with respect to equipment age and year of occurrence. Sparsity highlighted.

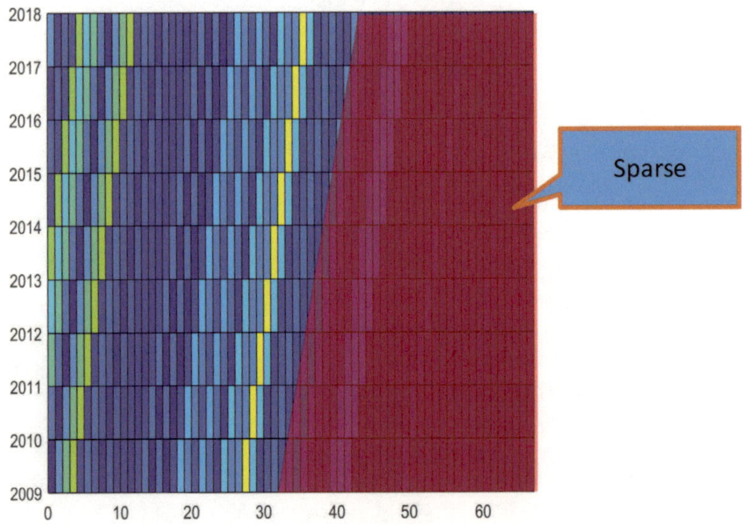

Figure 5.
Number of equipment with respect to equipment age. Sparsity highlighted.

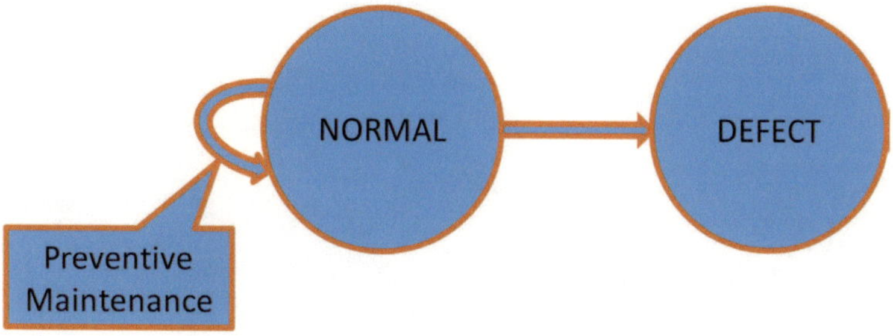

Figure 6.
State transition functions (I).

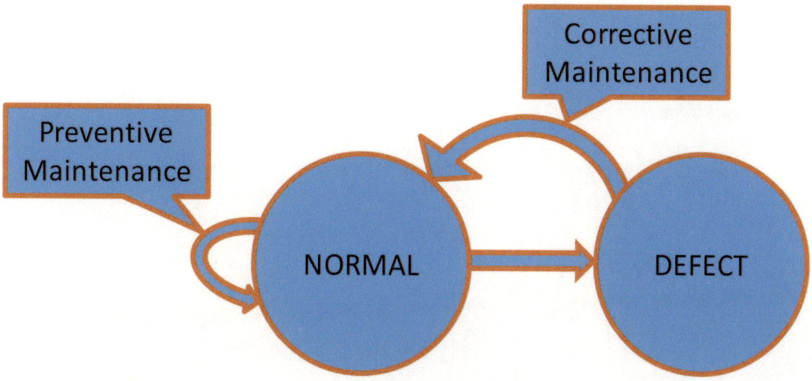

Figure 7.
State transition functions (II).

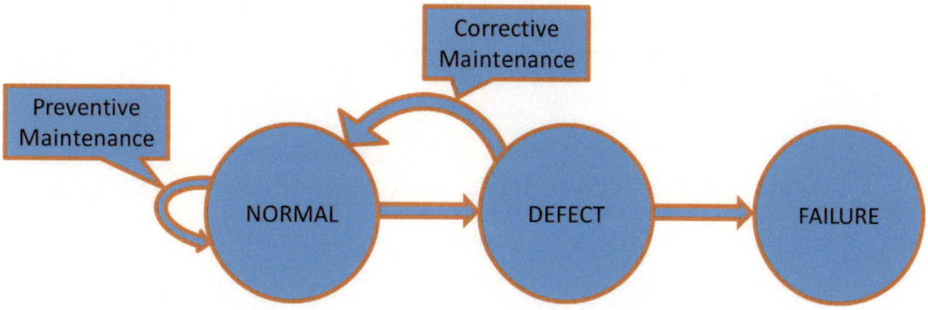

Figure 8.
State transition functions (III).

In the case of transformers, the **FAILURE** state can be divided into two others, i.e., **Internal Failure** and **External failure**, as shown in **Figure 9**.

The approximation of the defect rate function should take into account the quality of the available information. The quality of the information depends on the number of maintenance records available and the number of devices [6]. The relative quality of the data can be seen in **Figure 10**.

A similar approximation can be made for the failure rate, as depicted in **Figure 11**, which is modeled by means of a power type function. **Figure 12** also shows the failures involving peripheral elements and the active parts of transformers.

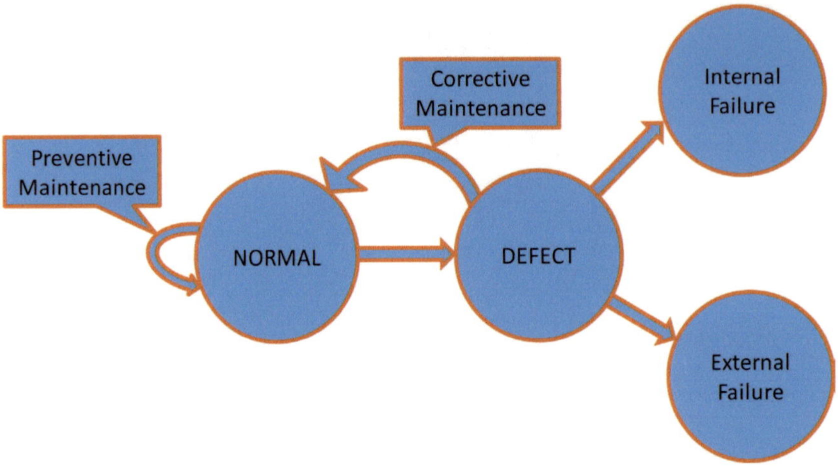

Figure 9.
State transition functions (IV).

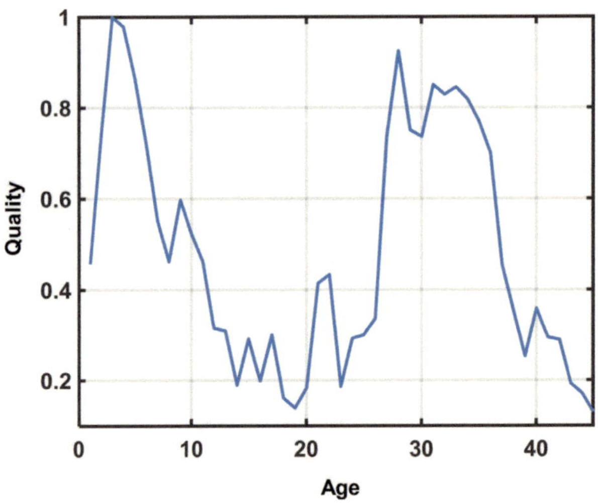

Figure 10.
Relative quality of the available data as a function of the asset age.

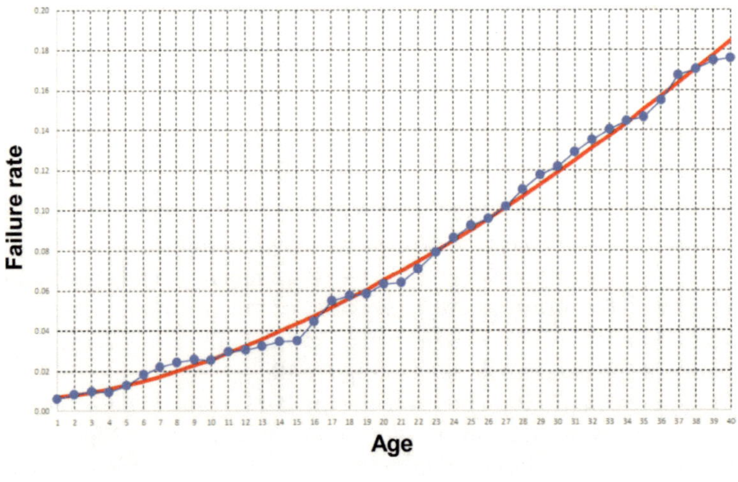

Figure 11.
Approximation of the failure rate.

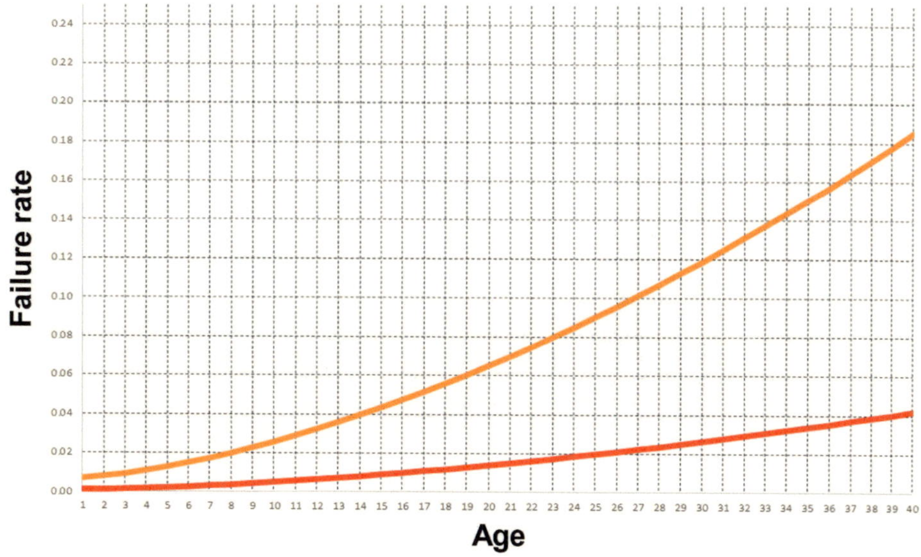

Figure 12.
Failures for peripheral (orange) elements and active (red) parts.

3. Business intelligence interface

All intelligence embedded in the AMSS systems was implemented using the concept of Business Intelligence, which uses graphical reports to represent data. The software was divided into nine pages, with several graphical analyses each. The first page can be seen in **Figure 13**.

In this screen, the user can choose the **Substation**, the **Equipment** type, and also the **Voltage** class, among **729** available equipment in **4843** available corrective maintenance records, as exemplified in **Figure 14**.

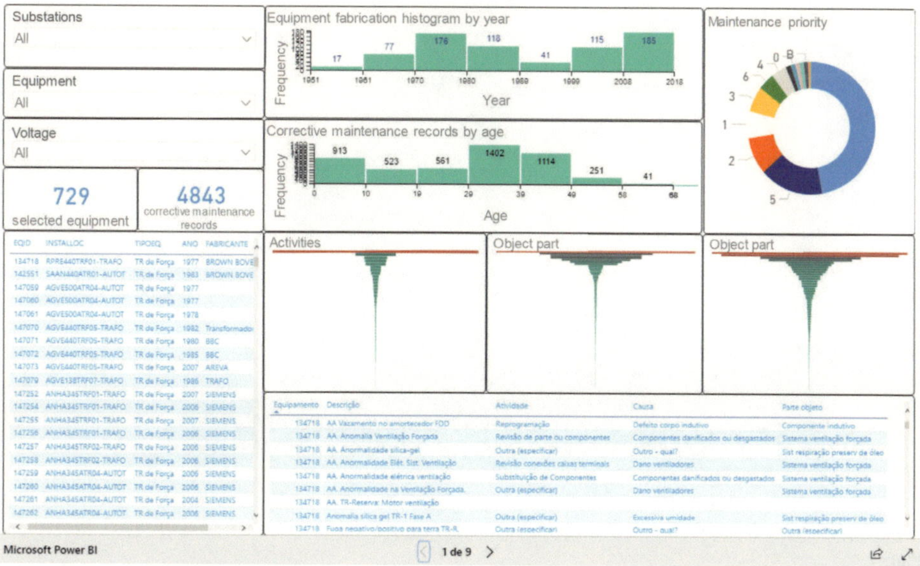

Figure 13.
Business intelligence Interface page 1.

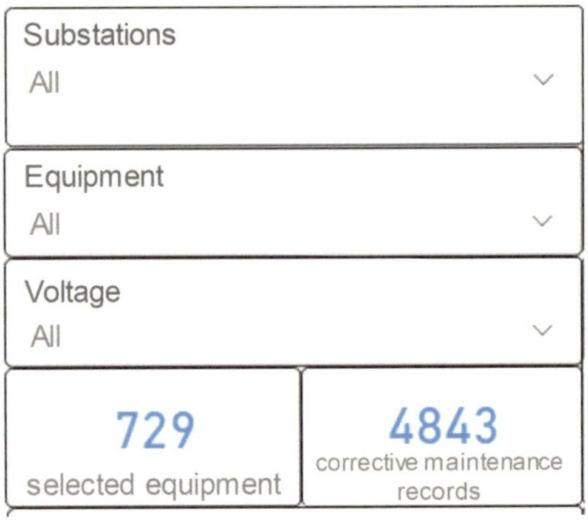

Figure 14.
Selecting substations, equipment, and voltage class detail from page 1.

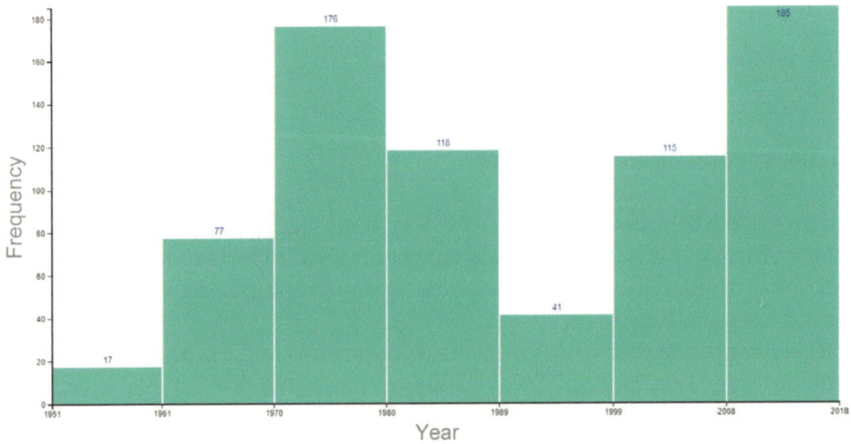

Figure 15.
Equipment fabrication histogram by year (page 1).

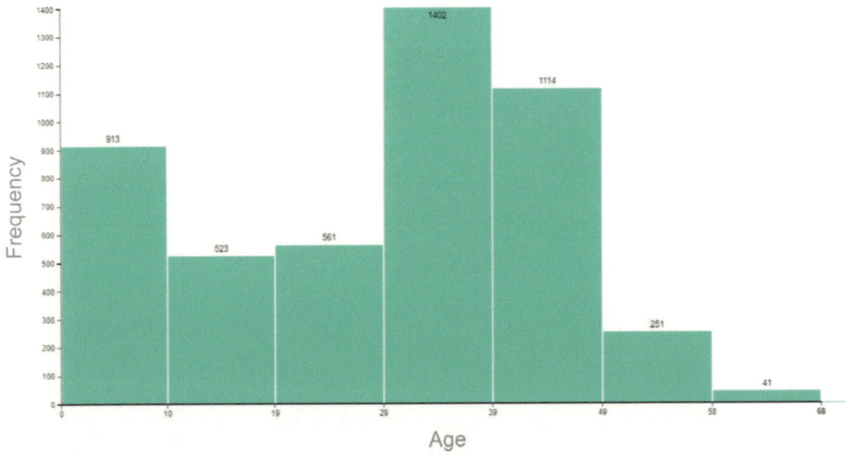

Figure 16.
Corrective maintenance records by age (page 1).

In **Figure 13**, it is still possible to visualize, in detail, the histogram of the manufacturing year of the e**quipment**, as well as the **Corrective maintenance records by age**, detailed in **Figures 15** and **16**, respectively.

The **Maintenance priority** is detailed in **Figure 17**.

Page 2 displays graphical reports involving all corrective maintenance records for all devices, as can be shown in **Figure 18**.

In Page 3 of the AMSS system, **Figure 19**, it is possible to see the correlations of corrective maintenance records with the age of the assets, which are shown in **Figures 20** and **21**.

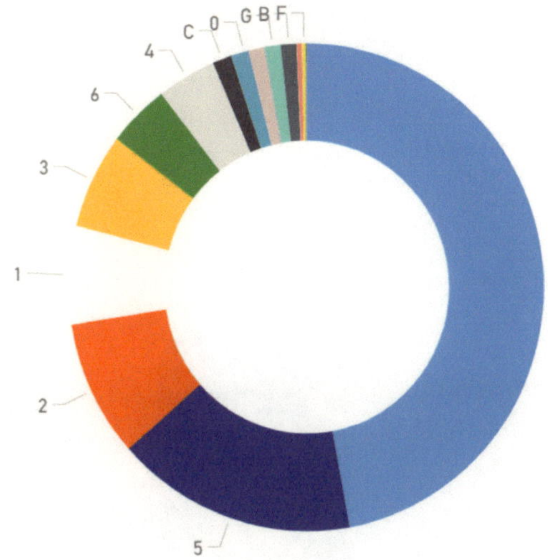

Figure 17.
Maintenance priority (page 1).

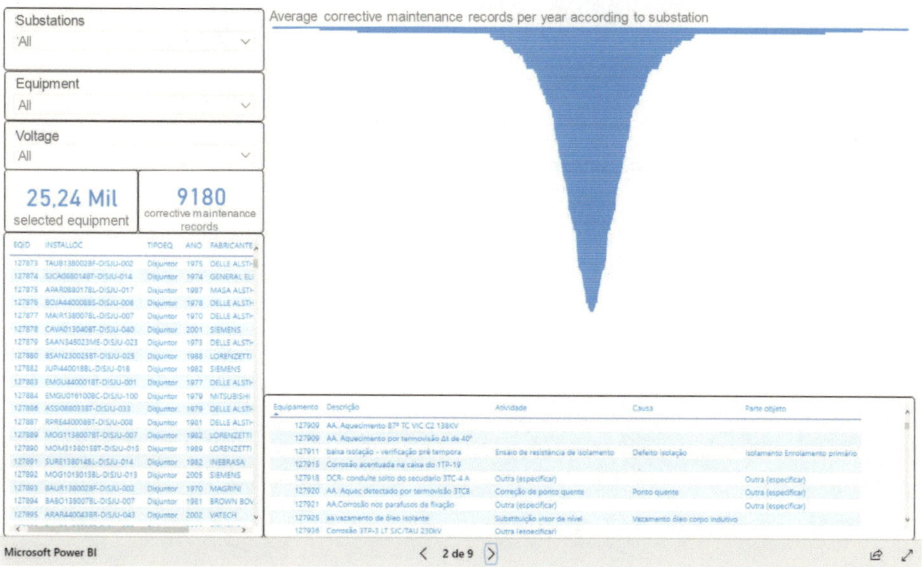

Figure 18.
Business intelligence Interface page 2.

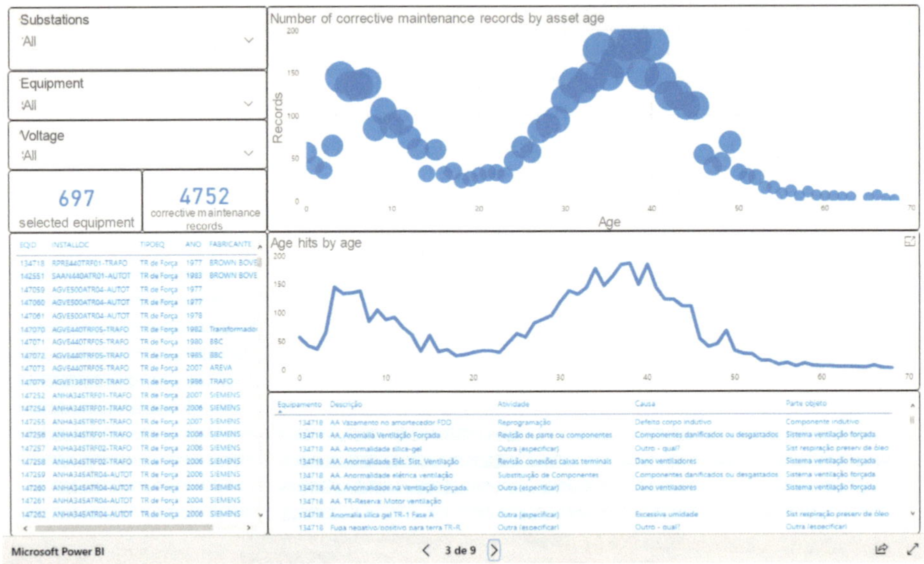

Figure 19.
Business intelligence Interface page 3.

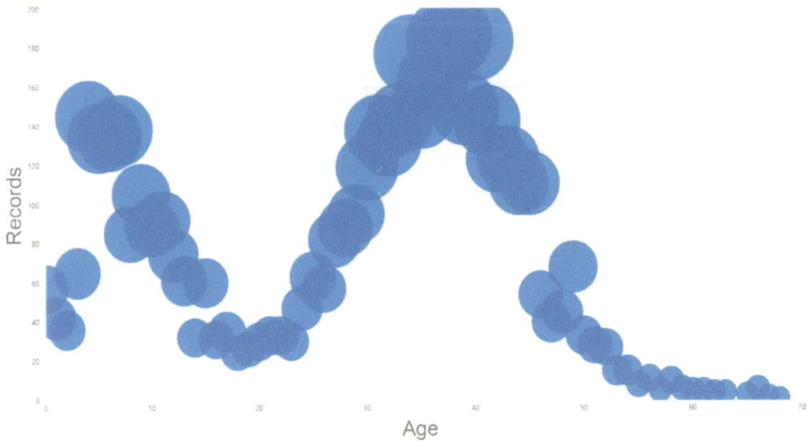

Figure 20.
Detailed number of corrective maintenance records by asset age from page 3.

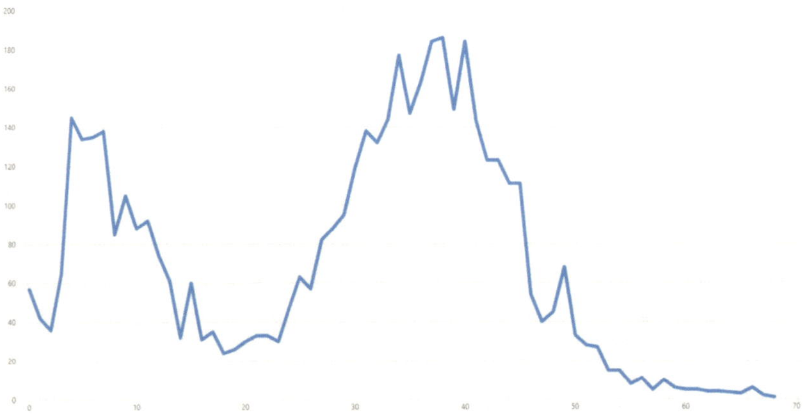

Figure 21.
Detailed age count by age from page 3.

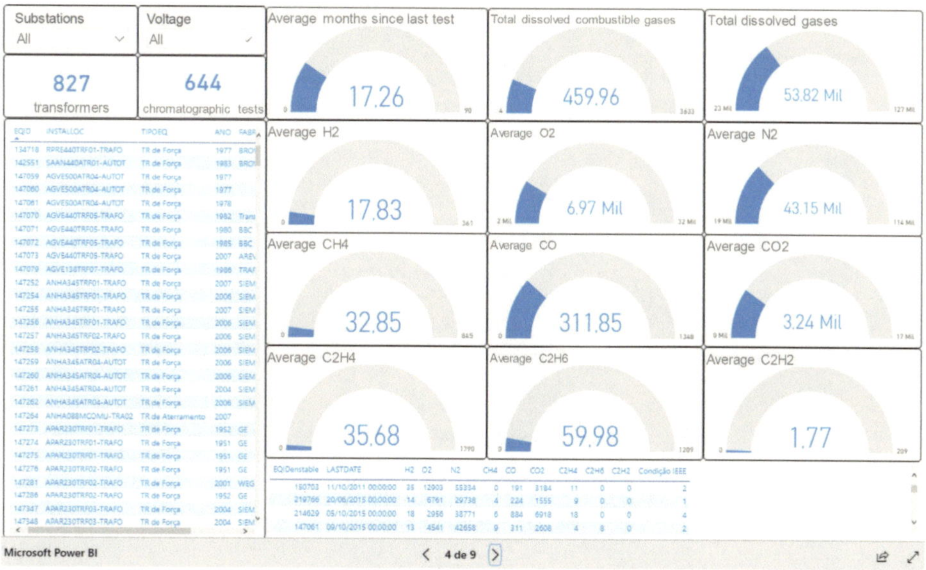

Figure 22.
Business intelligence Interface page 4.

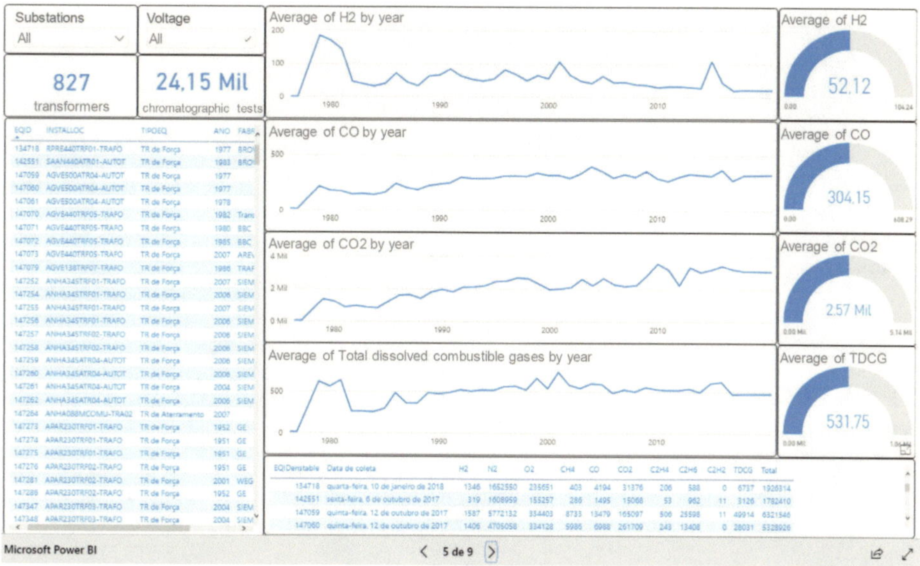

Figure 23.
Business intelligence Interface page 5.

Page 4 presents the user with information from chromatographic assays for power transformers that can be individually selected, as can be seen in **Figure 22**.

Pages 5 and 6 still address information from chromatography tests, showing the gas emission evolution over the years, as well as the correlation between the tests and the corrective maintenance history for each asset, as seen in **Figures 23** and **24**, respectively.

The emission evolution of H_2, CO, CO_2, total dissolved combustible gases, CH_4, C_2H_4, C_2H_6, and C_2H_2 can be seen in details in **Figures 25–32**.

Page 7 shows the behavior of the dissolved gases as a function of time. The user can select a particular asset or analyze the evolution globally, this is, for all assets.

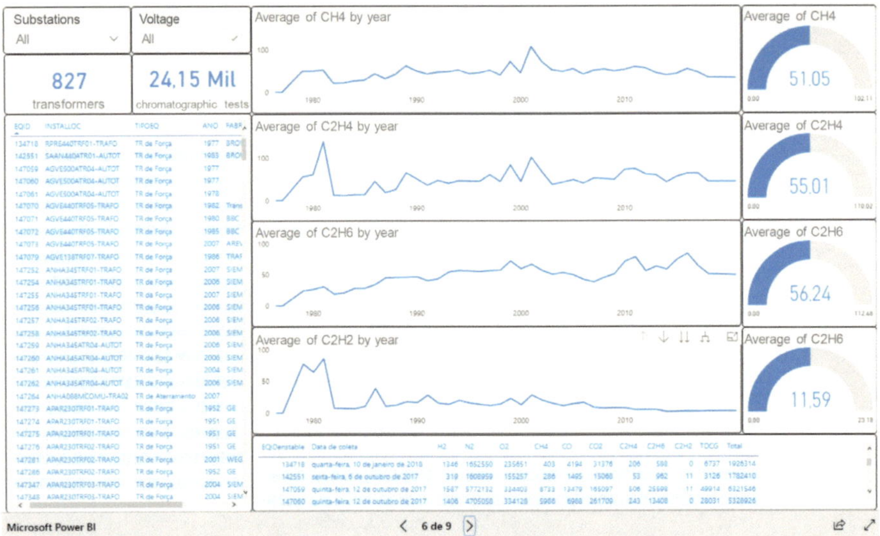

Figure 24.
Business intelligence Interface page 6.

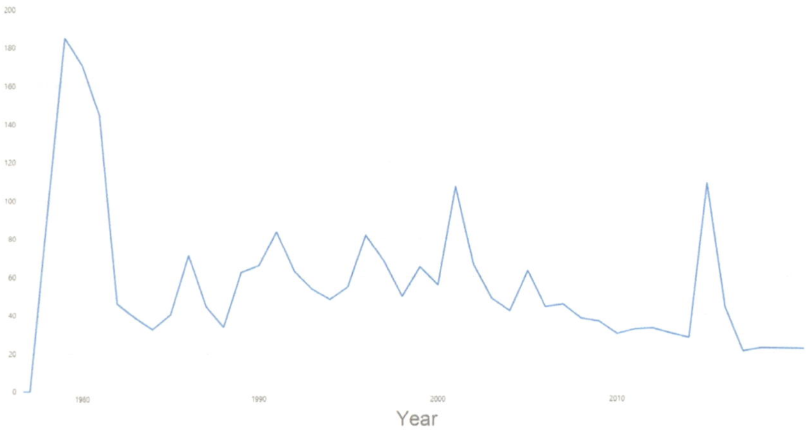

Figure 25.
Detailed H_2 emission evolution from page 5.

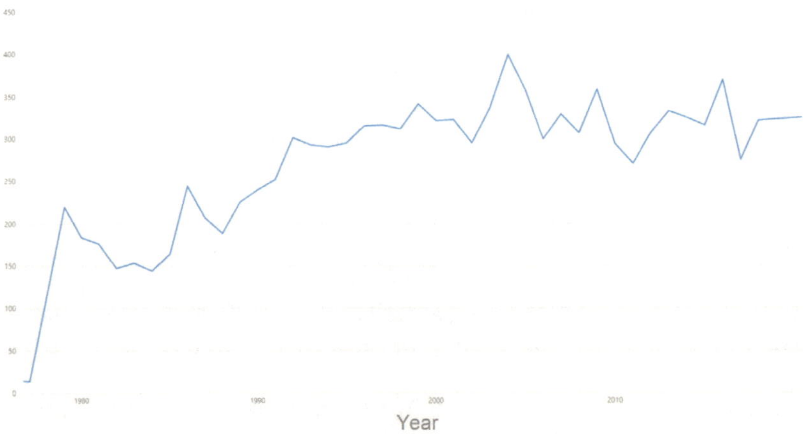

Figure 26.
Detailed CO emission evolution from page 5.

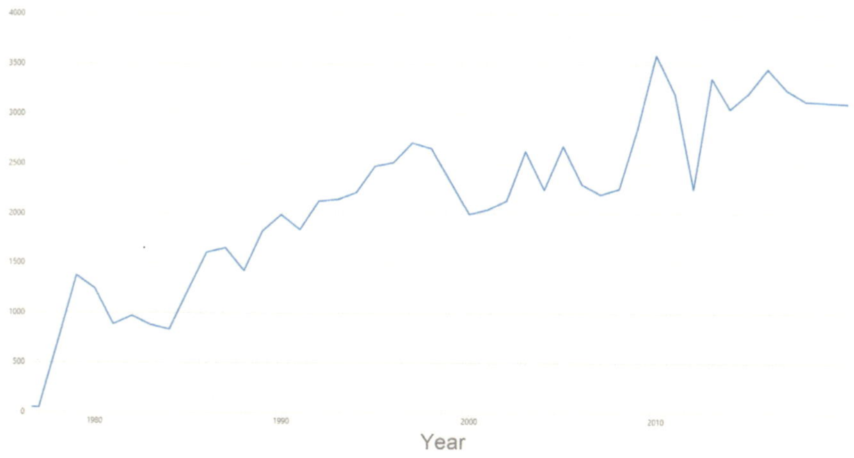

Figure 27.
Detailed CO_2 emission evolution from page 5.

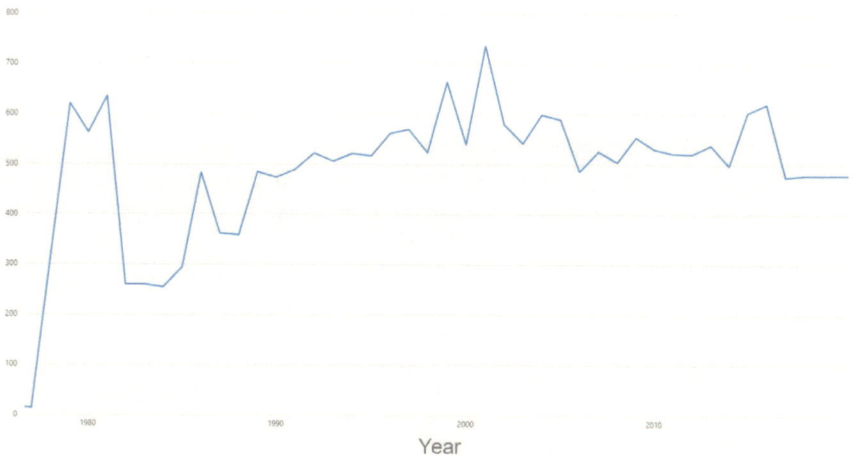

Figure 28.
Detailed TDCG emission evolution from page 5.

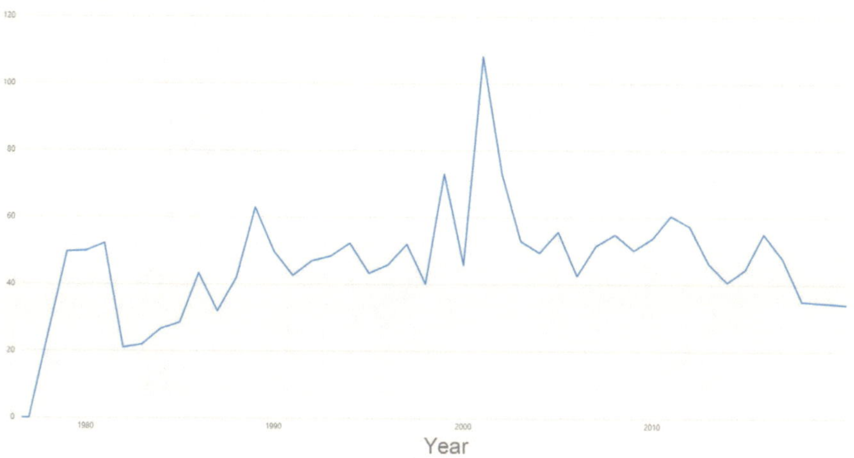

Figure 29.
Detailed CH_4 emission evolution from page 6.

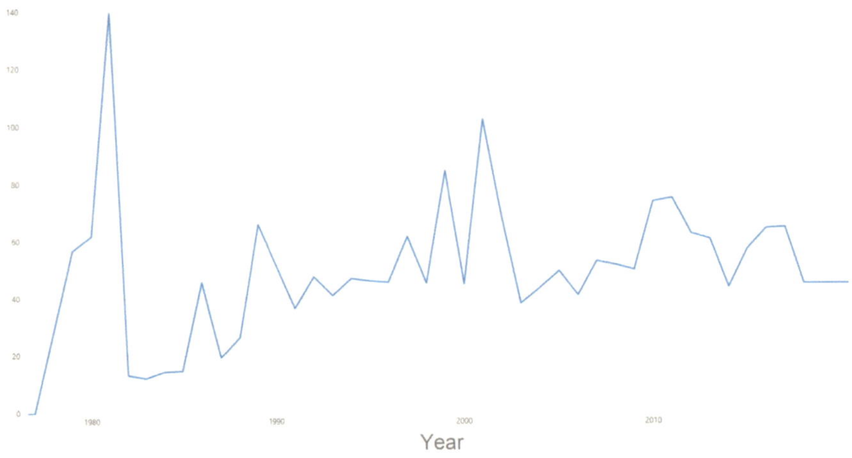

Figure 30.
Detailed C_2H_4 emission evolution from page 6.

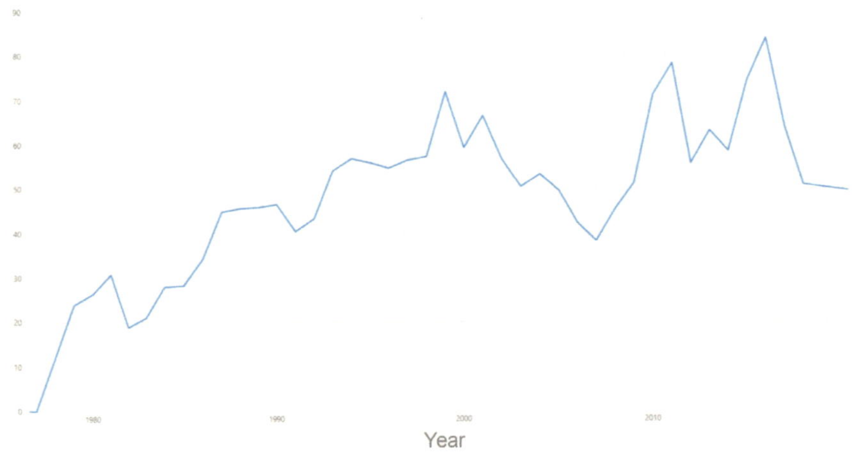

Figure 31.
Detailed C_2H_6 emission evolution from page 6.

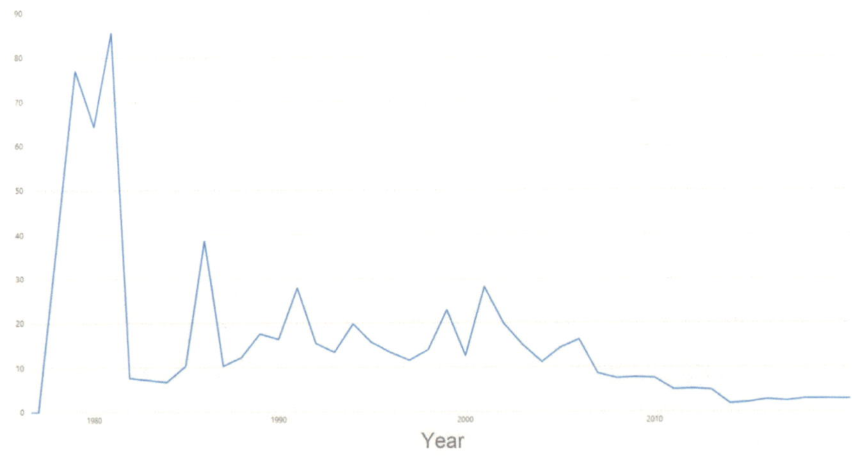

Figure 32.
Detailed C_2H_2 emission evolution from page 6.

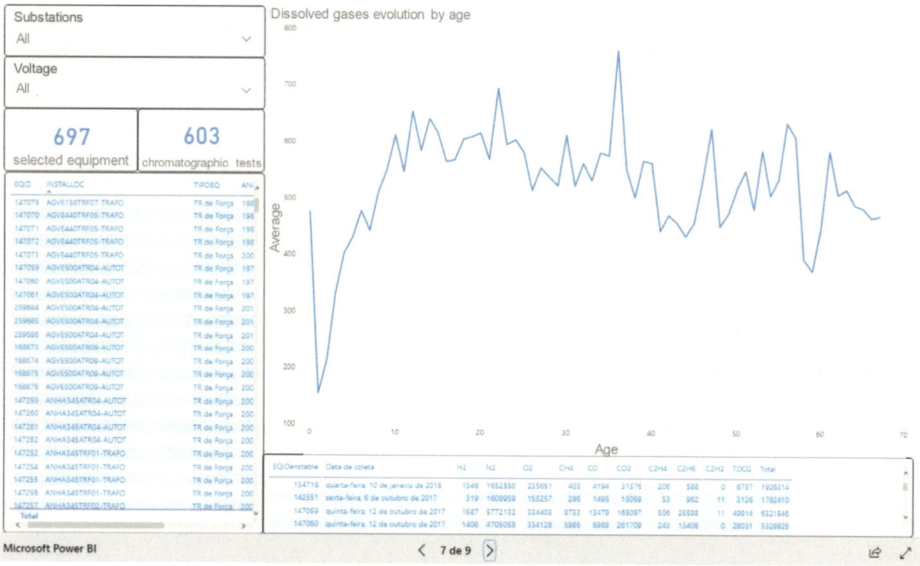

Figure 33.
Business intelligence Interface page 7.

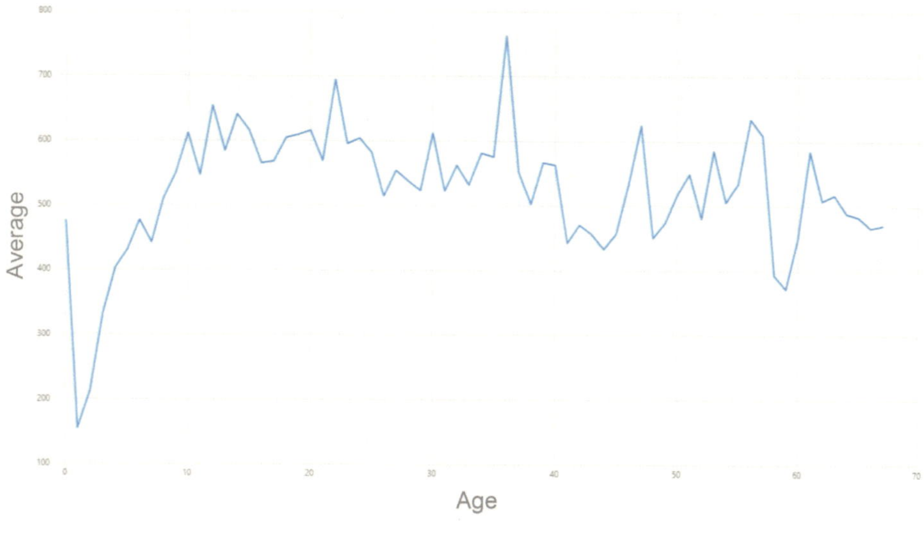

Figure 34.
Detailed dissolved gases evolution from page 7.

This behavior is shown in **Figure 33**. The detailed evolution of the dissolved gases as a function of the age of the assets can be seen in **Figure 34**.

Page 8 presents the integrated analyses involving the relationship between corrective maintenance records and chromatographic tests, as exemplified in **Figure 35**. **Figure 36** shows the trend line of maintenance records per year as a function of the age of the asset.

Finally, Page 9, which is shown in **Figure 37**, performs an integrated analysis of the health of an asset, along with its maintenance history.

It can be seen from **Figure 37** that the circles represent the transformers, and the diameter of each circle is related to the corrective maintenance rate per year to which this asset is subjected. The ordinate axis represents the current condition that this asset is in.

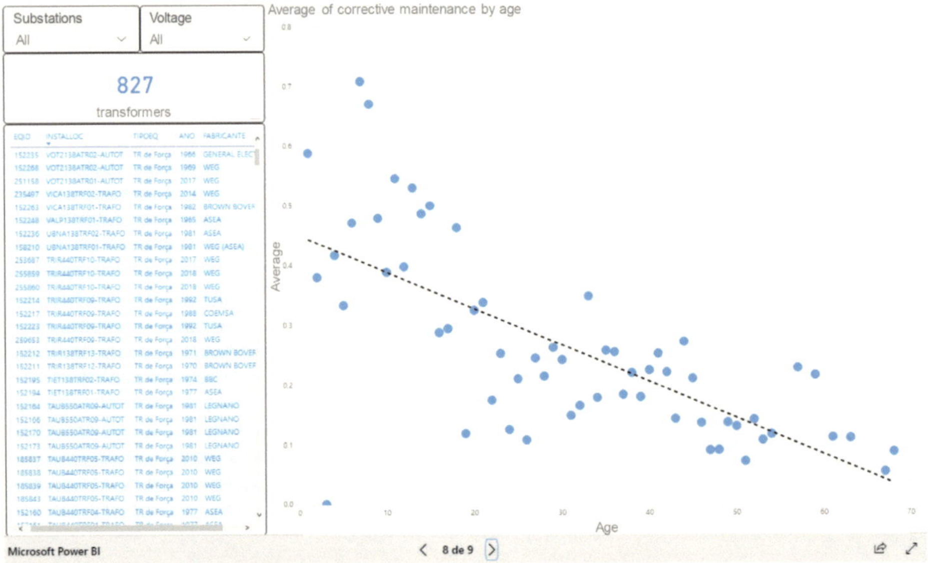

Figure 35.
Business intelligence Interface page 8.

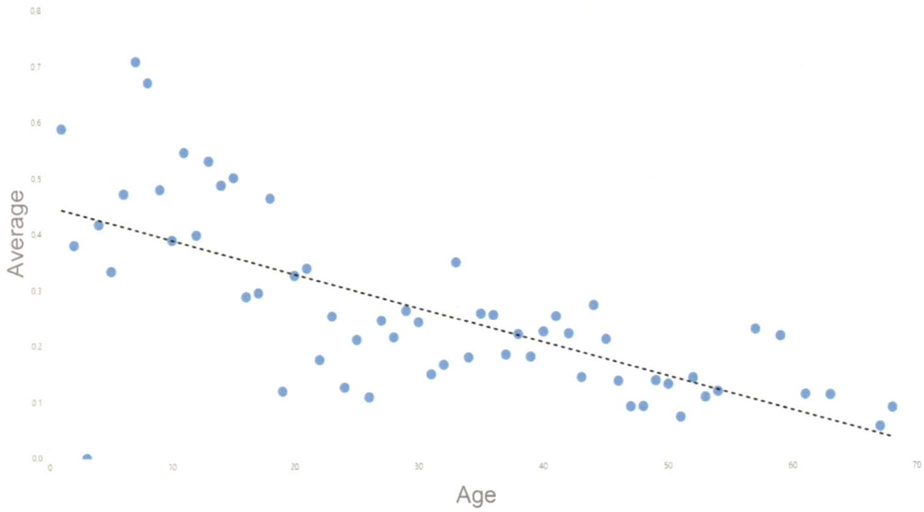

Figure 36.
Detailed trend line for corrective maintenance from page 8.

4. Conclusions

Asset management in the power sector, especially in transmission systems, has been driving the development of increasingly efficient feature extraction tools.

This chapter has introduced an integrated method based on business intelligence that analyzes data and failure rates in order to assist decision-making. The computational system takes into account technical information from the assets, data from chromatographic tests, as well as standard information regarding the operative condition of the assets, especially power transformers.

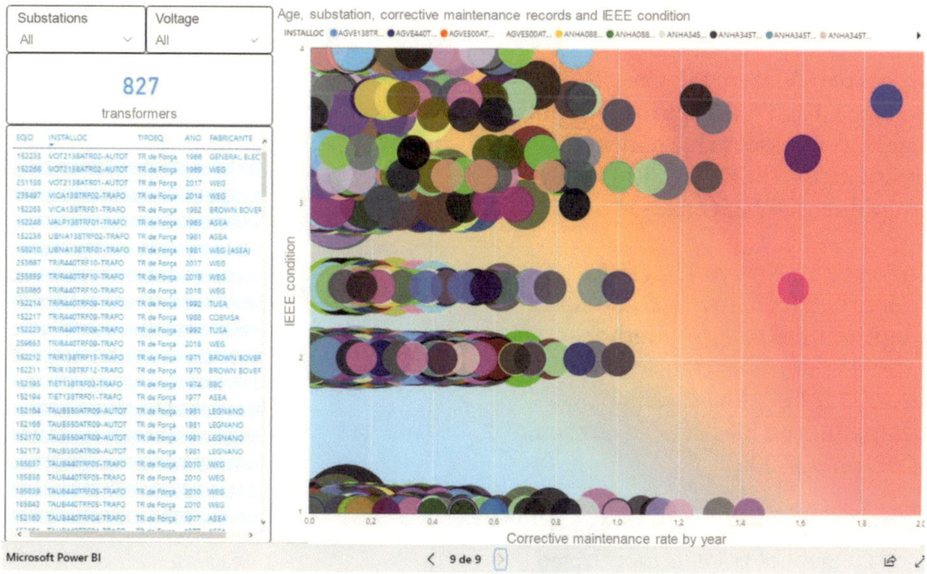

Figure 37.
Business intelligence Interface page 9.

Acknowledgements

The authors thank the ANEEL R & D Program contract number PD-0068-0037/2016.

Author details

Danilo Spatti[1]*, Luisa H.B. Liboni[2], Marcel Araújo[3], Renato Bossolan[4] and Bruno Vitti[4]

1 University of São Paulo, São Carlos, Brazil

2 Federal Institute of Education, Science, and Technology of São Paulo, Sertãozinho, Brazil

3 Federal Rural University of Pernambuco, Recife, Brazil

4 São Paulo State Electric Power Transmission Company, São Paulo, Brazil

*Address all correspondence to: spatti@icmc.usp.br

References

[1] Jahromi A, Piercy R, Cress S, Service J, Fan W. An approach to power transformer asset management using health index. IEEE Electrical Insulation Magazine. 2009;**25**(2):20-34

[2] Nemeth B, Benyo T, Jager A, Csepes G, Woynarovich G. Complex diagnostic methods for lifetime extension of power transformers. In: 2008 IEEE ISEI—International Symposium on Electrical Insulation. 2008. pp. 132-135

[3] Zhang X, Gockenbach E. Asset-management of transformers based on condition monitoring and standard diagnosis. IEEE Electrical Insulation Magazine. 2008;**24**(4):26-40

[4] Coble JB. Merging data sources to predict remaining useful life, an automated method to identify prognostic parameters (University of Tennessee PhD dissertation). 2010

[5] Heo JH, Kim MK, Park GP, Yoon YT, Park JK, Lee SS, et al. A reliability-centered approach to an optimal maintenance strategy in transmission systems using a genetic algorithm. IEEE Transactions on Power Delivery. 2011;**26**(4):2171-2179

[6] Sarchiz D, Bica D, Georgescu O. Mathematical model of reliability centered, maintenance (RCM). Power transmission and distribution networks applications. In: 2009 IEEE PowerTech. 2009. 4 p